T0186346

# GEOGRAPHIC INFORMATION SYSTEMS IN OCEANOGRAPHY AND FISHERIES

# GEOGRAPHIC INFORMATION SYSTEMS IN OCEANOGRAPHY AND FISHERIES

*Vasilis D. Valavanis*

**CRC Press**
Taylor & Francis Group
Boca Raton London New York

CRC Press is an imprint of the
Taylor & Francis Group, an **informa** business
A TAYLOR & FRANCIS BOOK

CRC Press
Taylor & Francis Group
6000 Broken Sound Parkway NW, Suite 300
Boca Raton, FL 33487-2742

First issued in paperback 2019

© 2002 by Vasilis D. Valavanis
CRC Press is an imprint of Taylor & Francis Group, an Informa business

No claim to original U.S. Government works

ISBN-13: 978-0-415-28463-9 (hbk)
ISBN-13: 978-0-367-39613-8 (pbk)

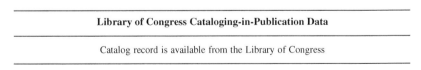

**Library of Congress Cataloging-in-Publication Data**

Catalog record is available from the Library of Congress

**Visit the Taylor & Francis Web site at
http://www.taylorandfrancis.com**

**and the CRC Press Web site at
http://www.crcpress.com**

TO CHRYSOULA, DIMITRA AND ANASTASIA

# CONTENTS

# COLOUR PLATES

*(Between pages 108 and 109)*

# FIGURES

# TABLES

# FOREWORD

I had the pleasure of reading the manuscript for this book whilst holidaying on the southern coast of the island of Hvar, one of the Dalmatian islands in the Adriatic Sea. Whilst there I reflected on the fact that the seas in this immediate area were rather like the holiday itself – they were an escape from the reality that existed through much of time and space. Thus, the waters were sparklingly clean and short spells of snorkelling revealed an abundant array of numerous fish. How different from the situation faced by an increasing proportion of the world's fisheries and oceans!

Vasilis's book serves the fundamental purpose of addressing some of the problems that almost universally beset oceans and their fisheries. It does not attempt to do this in the normal way by undertaking a 'problem to solution' synthesis. Thus, the book does not attempt to summarise the global demise of fin-fish stocks, or spell out the myriad problems and complexities relating to fisheries or ocean management, and then to make tentative solutions. This book goes a step further. It says – 'Look we know about the problems. This is how we can best bring information technology to bear upon these challenges. This is how we might carry out the exhaustive analyses that are necessary when considering problems in a complex marine milieu'. The intention of the book is therefore to illustrate the potential for GIS-based analyses, and to show the main technical and methodological means of confronting the demise of oceanic biological systems. In attempting this difficult task, the book achieves its goals in a very positive way. Given the plight of many marine systems, this is indeed a very timely achievement.

Over the last two decades, there has been increasing recognition that problems in fisheries and related marine areas are nearly all manifest in the spatiotemporal domain. Putting it somewhat simply, much of the marine biological environment is out of equilibrium. We know that the reasons for this are imbedded in a miscellany of factors concerned with poor environmental management; a certain slovenliness in introducing or operating appropriate management techniques; species overfishing allied to excessive fishing capacity; and a simple lack of knowledge on means of optimising productivity in fisheries and other marine ecosystems. Either singly or collectively these reasons are manifest in biological systems that cannot be maintained and are thus in decline. However, to confront this challenge, the use of Geographical Information Systems (GIS), allied to other spatial management technologies, have emerged as a potential saviour. Thus, here is a set of systems that if carefully and judiciously applied, clearly have the potential to go a long way towards solving many of the space-related problems. Perhaps, they will provide the extra impetus needed to reverse the worrying marine biological trends.

An examination of the 'Contents' pages of this book gives an insight into the extraordinary breadth of subject matter that we must be concerned with. The oceans are truly complex places. They exhibit an almost infinite range of variables and processes that themselves might be integrated in endless combinations. If we are to examine problems in the marine environment, it is necessary to partition this environment into manageable, conceptually based classes or categories. Vasilis had made this very clear, and the reader can easily progress from one identifiable variable or process to another. Yet all the time linkages are stressed, and the reader is left in no doubt that variables or processes cannot be measured and studied in isolation.

Overcoming the enormously complex marine-based problems to which GIS work is being directed will not be easy, and it is likely that new problems will materialise at a rate similar to that at which existing problems will be overcome. The difficulties of applying GIS in the marine sphere greatly exceed those encountered with terrestrial applications. Here, the variables mapped are static and are frequently permanent. But in the marine realm almost everything moves, including the milieu itself! And most movements are chaotic and unpredictable. This leads to two major considerations that terrestrial GIS seldom must face: (i) how frequently should the variables or processes be mapped; and (ii) the resolution at which any mapping or data gathering should be carried out. These considerations are not easily resolved, and when they are, there will not be universal answers. Another complexity faced by marine GIS applications is that of operating in a 3D environment. The added dimension has enormous consequences for data volumes and data storage, for spatial analyses and for more basic considerations related to visualisation of GIS output. Although progress is being made in applications of 3D GIS, these are mostly where they are used for subsurface, soil or geology structural mapping, applications where the moving milieu is not an additional problem. Whilst this book may not provide answers to many of the additional problems related to working with marine GIS's, it will certainly steer us towards much of the work that is being carried out here.

Probably the greatest strength of Vasilis's work is the huge range of appropriate research material that he has gathered together. I have been involved in the area of 'fisheries GIS' for nearly two decades, and I try to keep abreast with developments, but I can honestly say that I had no idea that such an extraordinary range of work was being undertaken. Thus, we are provided with a very detailed synopsis of all the latest applications to marine sciences, not only in GIS but also in associated fields such as remote sensing and acoustic sonar. Additionally, data sources are explored, variations of appropriate software and hardware are examined, and potential methods of adopting GIS for management purposes are discussed. Simply finding all of this source material was a notable achievement, and if the reader wished to get added utility from the book, he/she might approach Vasilis for insights into his information search mechanisms!

As I have intimated earlier, this book is both timely and original. It does not specifically complement other books; instead it is unique and it greatly adds to our store of information on what is taking place in the spheres of fisheries and oceanography, particularly in the subfields of spatial and temporal analyses. For anyone just commencing work in this subject area, the work is invaluable. For those of us who have been working in this field for even a short period, and who have an awareness of how great are the scope and magnitude of the problems, then the book will be a guide to additional and potential problem-solving sources. I truly believe that this work will make a significant contribution to problem resolution in both the fisheries and oceanographic environments.

Geoff Meaden
Canterbury, UK
November, 2001

# PREFACE

This book was conceived in the summer of 2000. At that time, a European Commission funded, 3-year project on cephalopod resource dynamics was in its completion and the Millennium Cephalopod Conference of the Cephalopod International Advisory Committee (CIAC) was in progress at the University of Aberdeen, Scotland. One of the CIAC Conference satellite meetings was the Geographic Information Systems (GIS) and Fisheries Workshop, where GIS scientists and marine biologists from all over the world exchanged ideas on GIS implementation in cephalopod fisheries. The need for information-based fisheries management proposals was underlined and progress of marine and fisheries GIS developments to this goal was evident.

From the start, this book was organised around the need for information-based marine and fisheries management and especially on how GIS can contribute and facilitate these processes. GIS technology, as a new technology, is continually under development progressing at a rapid pace. In the book, many oceanographic and fisheries GIS applications are reviewed, applications that present a high variety of methods and sophisticated approaches. For example, in oceanography, GIS suggests methods for the mapping and measurement of major ocean processes that greatly affect the state of marine environment while in fisheries, GIS provides a suitable framework for the facilitation of the complex fisheries management process. The contents of this book present general GIS issues through specific marine and fisheries applications providing also related GIS routines. The book is organised into four main sections: The first part describes the main components of a marine GIS, the relation of GIS with similar technologies, conceptual issues on marine spatial thinking, models of marine GIS development with emphasis to the essential goal of any GIS, that of generating information-based management proposals. The second part presents the main sampling methods and online sources of spatially referenced oceanographic data and covers application examples on how GIS contribute to the mapping of certain oceanographic phenomena (upwelling, front, gyre, etc.), deep ocean environments and other oceanic studies. The third part presents various fisheries monitoring methods and online sources of spatially referenced fisheries data and covers fisheries application examples revealing how GIS contribute to the identification of spatiotemporal components of marine species population dynamics (spawning grounds, essential habitats, migration corridors, etc.). Both parts on GIS in oceanography and fisheries examine an extensive number of applications. The purpose of this examination is to present the many different areas and variety of ways GIS are used in these fields and provide ideas for further GIS developments. Finally, the fourth part (Annex I and II) presents GIS technical issues by listing the marine GIS routines for a wide array of GIS tasks (data downloading and GIS database design, data analysis, integration, output, and system interfacing).

It is anticipated that the relevance of the book will be such that anyone with interests in marine GIS development, physical and biological oceanography, fisheries and information-based proposals for marine resource management will find it useful. The aim of the idea of producing a book that examines general marine GIS issues through a great number of reviewed applications and GIS routine presentation is to inspire others to produce further potential developments in the increasingly developing and highly related fields of oceanographic and Fisheries GIS. Without doubt, such applications offer suitable tools for information-based management of marine resources and provide a fascinating way to study the marine environment.

The author acknowledges with gratitude the support in various levels of Stratis Georgakarakos, Argiris Kapandagakis, John Laurijsen, John Haralabous, Panos Drakopoulos, Christos Arvanitidis, Kostas Dounas, Katerine Siakavara, Antonis Magoulas, George Kotoulas, Andrew Banks (Institute of Marine Biology of Crete, Greece), Tassos Eleftheriou, (University of Crete, Greece), Drosos Koutsoubas (University of the Aegean, Greece), Peter Boyle, Graham Pierce, Jianjun Wang (University of Aberdeen, UK), Paul Roadhouse, Phil Trathan, (British Antarctic Survey, UK), Arthur Cracknell (University of Dundee), Daniel Brackett, Scott Smith (University of Florida, USA), Ge Sun (North Carolina State University, USA), Darius Bartlett (Cork University, Ireland), Dawn Wright (Oregon State University, USA), Joao Pereira (Instituto de Investigacao das Pescas e do Mar, Portugal), Gildas Lecorre (Institut Francais de Recherche pour l' Exploration de la Mer, France), Eduardo Balguerias (Centro Oceanografico de Canarias, Spain) and Vincent Dennis, Jean-Paul Robin (Universite de Caen, France). Author's cooperation among these colleagues either in various international and national projects or for invaluable discussion and advice on the organisation of this book was of great value for the initiation, compilation and completion of this research.

The author highly acknowledges with obligation the following people, who willingly contributed their work by presenting their latest marine and fisheries GIS applications greatly enhancing the contents of this book:

- Fabio Carocci, Jacek Majkowski and Francoise Schatto (Food and Agricultural Organisation of the United Nations – FAO, Rome, Italy)
- Falk Huettmann (Simon Fraser University, Canada)

- Rick Lathrop, Phoebe Zhang, and Jen Gregg (Rutgers University, USA)
- Geoffrey Matthews (National Marine Fisheries Service, USA)
- Helena Molina-Urena (University of Miami, USA)
- Kimberly Murray (Woods Hole Oceanographic Institute, USA)
- Juan-Pablo Pertierra (Commission of European Communities, Belgium)
- Terry Peterson (MicroImages, Inc., USA)
- Teresa Pina (University of the Algavre, Portugal)
- Mitchell Roffer (Roffer's Ocean Fishing Forecasting Service, Inc., Miami, FL, USA)
- Claire Waluba (British Antarctic Survey, UK)

The author is obliged to Mrs Margaret Eleftheriou for her thorough and constructive reading of Chapter 1 of the manuscript and greatly acknowledges the four anonymous referees of the book's proposal. The author is highly obliged to Geoff J. Meaden (Canterbury Christ Church College, UK) for his kind overall help and critical and constructive reading of the manuscript.

CHAPTER ONE

# Marine Geographic Information Systems

## 1.1 INTRODUCTION

Geographic Information Systems (GIS) were 'born' on land; they are around 35 to 40 years old, but only about 15 years ago did they 'migrate' to the sea. In this process '. . . as fish adapted to the terrestrial environment by evolving to amphibians, so GIS must adapt to the marine environment by evolving and adaptation' (Goodchild 2000). The domain of GIS concerns georeferenced data, plus integration and analysis procedures, that function to transform the raw data into meaningful information that can support management decisions. In any environmental GIS, after defining the nature of the problem, the initial activity will be to measure aspects of the variable or natural process including both spatial and temporal perspectives. Variables or processes will have three types of properties that need recording: (1) features; (2) attributes; and (3) relationships. GIS will have the ability to store and access digital details of these measurements from a computer database. Then, measurements will be linked to features on a digital map. Aspects of the features will be able to be digitally mapped as points, lines, and polygons (vector) as well as pixels and voxels (raster). The analysis of collected measurements as well as the application of numerical manipulations or modelling algorithms may produce additional data. The combined analysis of several datasets in a GIS environment provides meaningful information for natural processes and it is the core of the GIS technique. The depiction of the analysed data in some type of display (maps, graphs, lists, reports or summary statistics) provides for the communication media of GIS results or output.

Li and Saxena (1993) as well as Lockwood and Li (1995) described the important differences between marine and terrestrial GIS applications. The static, terrestrial-based GIS developments consist of certain functions such as overlaying, buffering, reclassification and Boolean operators. Terrestrial objects are very suitable for such operations and output results with a high accuracy. Marine problems have the characteristics of the fuzzy boundary, dynamics, and a full three dimensions. It is not completely suitable for the land-based GIS to apply fully in the marine environment. In the marine context, GIS development enters into a highly dynamic environment where almost everything moves or changes. Marine GIS is called upon to describe the intimate relations among the wind and sea currents that trigger certain oceanographic processes and explain the impact of these processes to the behaviour of marine organisms, taking species biology and ecology into consideration, as well.

Wright and Bartlett (2000) identified the important contribution of GIS in coastal and oceanographic research by opening new ways of georeferenced data processing. They underlined the migration of the early ocean GIS applications that were simple data collection and display tools to complex integrated modelling and visualisation tools. They also pointed out the primitive stage at which volumetric GIS analysis and 3D GIS visualisation is today, underlining that marine GIS must first adapt to the characteristics of the marine world and marine data and then output results that describe the dynamic relations among the various components of the marine environment.

Meaden (2000) identified three new major components that are added to fisheries GIS applications: (1) the vertical dimension; (2) the dynamics of marine processes (upwelling, gyres, fronts, etc.); and (3) the dynamics of marine objects (species populations). In order to adapt to its new environment, marine GIS must transfer into its computerised environment existing knowledge from many marine disciplines, such as marine biology, physical and biological oceanography and use data from related technologies, such as Remote Sensing (RS) and Global Positioning System (GPS). These data are invaluable for the successful development of a marine GIS. Here, marine satellite sensors (AVHRR, SeaWiFS, TOPEX/Poseidon, etc.) provide a considerable amount of datasets that describe the sea surface environment well. In addition, marine surveys provide data for the vertical plane (though restricted in area coverage), although 3D marine datasets do exist for large areas deriving from moored and drifting buoys and oceanographic models. Fisheries statistical, biological, and genetic data are important data sources for marine objects.

An innovative approach to the development of marine GIS applications in fisheries is the introduction of species life history data to GIS analysis (Valavanis *et al.* 2002). These data refer to species biology and ecology and are valuable results from biological and genetic research. Today, species life history data are organised in tables or reports by many fisheries agencies and organisations worldwide and often refer to species populations that occupy a certain geographic area. Information on species spawning preferences, migration habits, recruitment periods, and optimum living environmental conditions are suitable for marine GIS analysis. Through GIS integration with fisheries production, fishing areas, and environmental data, species life history outputs valuable information on species spawning grounds, essential habitats, aggregation areas, and migration corridors in spatial and temporal scales.

Marine GIS, as a general term, includes a wide area of applications. Depending on the nature of questions that a marine GIS is called upon to answer and on the spatial extent they cover, applications in this field may be categorised as coastal, oceanographic, and fisheries GIS, with a good deal of overlap among all the three main kinds of marine GIS. A coastal fisheries GIS dealing with how oceanographic processes, such as upwelling, affect fish populations and production is a common example of the overlapping of marine disciplines in marine GIS applications. In such applications, GIS developers are called to cooperate with scientists from a variety of disciplines in order to design an application in such a way that specific spatiotemporal questions could be answered. Thus, marine GIS development is a multidisciplinary procedure and scientists from many disciplines are invited to participate. Marine and fisheries biologists as well as physical and biological oceanographers participate in the development process and check GIS input, analysis, and output for accuracy (McGwire and Goodchild 1997).

The generation of information-based management proposals (decision aid tools) is one of the main goals of a marine GIS development, which adds policy makers to the GIS development process. The ability of GIS to map integrations among a variety of datasets is unique for the identification of conflicts between current management policies and marine objects dynamics. GIS contribute to the enhancement of natural resource monitoring and test the efficiency of currently applied management policies by presenting integrated products that describe the relations among biotic and abiotic resources and their current management schemes. The mapping of possible conflicts will lead to new information-based management proposals and schemes, which, depending on the case, will prove to be effective to the preservation, conservation, and sustainable management of marine resource dynamics.

In respect of monitoring ocean and fisheries dynamics and generating new information-based management schemes, GIS technology is closely related to several other technologies, such as GPS, RS, modelling, image processing, spatial statistics as well as the Internet (Figure 1.1). Cross-disciplinary integration is essential in gripping with the complexity of contemporary environmental problems (global climate change, human impacts on environment, mitigation of environmental hazards, etc.) using the substantial powers of computation for data analysis, process simulation, and decision aid (Clarke *et al.* 2000). Today, there is an increasing number of GIS consortia and organisations that promote communication among scientists from different disciplines towards establishing cross-disciplinary integration in GIS and facilitating the resolution of GIS-related issues worldwide (Table 1.1).

GPS technology is essential to data georeference. Attributing location to marine data is mandatory to GIS analysis and gives a meaning to the term 'geography' of the oceans. GIS constitutes the natural framework in which geospatial data should be stored and analysed, and it is GPS that provides that essential information of geolocation to remotely sensed and surveyed data. Earth Observation (EO) data, deriving from a variety of space borne and airborne sensors, constitute the main data source for marine GIS developments. Satellite sensors that sample ocean surface repetitively provide images of several ocean surface parameters, such as temperature distribution, chlorophyll concentration, wave height, and wind speed. These parameters are adequate for marine GIS integration for the study of the main coastal and oceanic processes such as upwelling, fronts, gyres and eddies that greatly affect biotic resources.

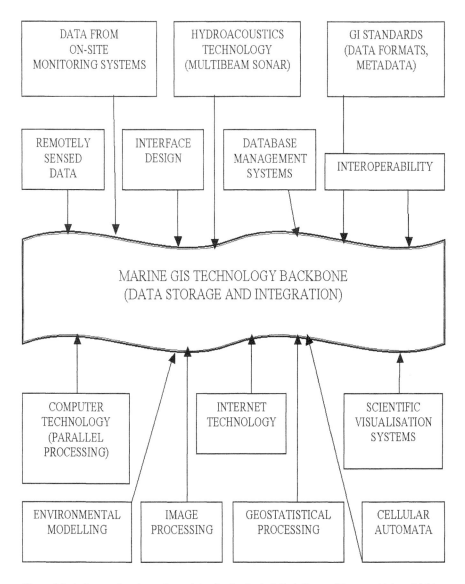

**Figure 1.1.** A diagram that shows the variety of technological disciplines and issues, which are highly associated with the main core of marine GIS technology.

**Table 1.1.** Major International and National GIS Consortia.

| NAME | SHORT INFORMATION | WEBSITE |
|---|---|---|
| AGI, Association for Geographic Information | An association of users in the private and public sectors, suppliers of software, hardware, data and services, consultants, and academics involved in research and teaching | http://www.agi.org.uk |
| AGILE, Association of Geographic Information Laboratories in Europe | To promote academic teaching and research on GIScience by representing the interests of those involved in GI-teaching and research at the national and the European level | http://romulus.arc.uniromal.it/Agile/Agile.html |
| CIESIN, Center for International Earth Science Information Network | An NGO to provide information that would help scientists, decision makers, and the general public better understand the changing world | http://www.ciesin.org |
| CORE, Consortium for Oceanographic Research and Education | CORE is organised under the US National Oceanographic Partnership Programme (NOPP) and it is an association of 66 oceanographic research institutions, universities, laboratories and aquaria representing the nucleus of US research and education about the ocean | http://core.ssc.erc.msstate.edu |
| GCDIS, Global Change Data and Information System | A Gateway to Global Change Data | http://globalchange.gov |
| GeoData Alliance | A non-profit organisation open to all individuals and institutions committed to using GI to improve the health of our communities, our economies, and the Earth | http://www.geoall.net |
| GIS-WRC, GIS Water Resources Consortium | A consortium for developing and implementing new GIS capabilities in water resources and design of a new GeoDatabase Model for Rivers and Watersheds | http://www.crwr.utexas.edu/giswr |
| MareNet, Network of Marine Research Institutions and Documents | A worldwide network providing a set of online information services, which enable marine scientists to communicate with the worldwide marine science community. | http://marenet.uni-oldenburg.de/MareNet |

| | | |
|---|---|---|
| MarLIN, Marine Life Information Network | An initiative of the UK Marine Biological Association in collaboration with major holders and users of marine biological data | http://www.marlin.ac.uk |
| MSC, Marine Science Consortium | A non-profit corporation dedicated to promote teaching and research in the marine sciences | http://www.msconsortium.org |
| NCGIA, National Center for Geographic Information and Analysis | An independent research consortium dedicated to basic research and education in GIScience and its related technologies, including GISystems | http://www.ncgia.ucsb.edu |
| NSDI, National Spatial Data Infrastructure | US infrastructure to reduce duplication of effort among agencies, improve quality and reduce costs related to GI, to make geographic data more accessible to the public, to increase the benefits of using available data, and to establish key partnerships among stakeholders | http://www.fgdc.gov/nsdi/nsdi.html |
| NSGIC, National States Geographic Information Council | An organisation of delegations of senior state GIS managers from across the US committed to efficient and effective government through the prudent adoption of information technology | http://www.nsgic.org/netframe.htm |
| Open GIS Consortium | A worldwide consortium of GI-related academic institutions and companies to deliver spatial interface specifications that are openly available for global use | http://www.opengis.org |
| UCGIS, University Consortium for Geographic Information Science | A non-profit organisation of US academic institutions dedicated to advancing our understanding of geographic processes and spatial relationships through improved theory, methods, technology, and data | http://www.ucgis.org |

Today, the integration of RS data within GIS is a routine task in marine GIS developments. The multidisciplinary RS data are used in GIS analysis in a variety of tasks and studies, such as global change (Asrar 1997), rectification and registration of satellite imagery (Ehlers 1997), change detection of marine processes (Jensen *et al.* 1997), and visualisation of these processes (Faust and Star 1997). In addition, several spatial analysis methods are used in RS and GIS analysis, which contribute to the continually evolving nature of GIS and RS integration. Spatial analysis for RS and GIS requires new and evolving techniques, such as data classification methods (e.g. artificial neural networks and fuzzy classification), geostatistical analysis (e.g. spatial interpolation and kriging methods), and spatial analysis and accuracy of classified data (Atkinson and Tate

1999). Designing and implementing software for spatial statistical analysis in GIS environments are already presented (e.g. Haining *et al.* 2000; Marble 2000) and several applications were proposed, for example, integration of S-PLUS and ArcView (Bao *et al.* 2000) and the use of R language (Bivand and Gebhardt 2000).

Another integration between similar disciplines is that of GIS and environmental modelling. Goodchild *et al.* (1993) reviewed the status of this integration while Johnston *et al.* (1996) discussed the use of GIS to the study of dynamic ecological processes. Today, modelled output is routinely integrated into GIS applications, particularly for the study and forecasting of large-scale oceanographic and atmospheric processes. An existing issue in this integration is that of data uncertainty. Specifically focused on the fusion of activities among GIS analysis, RS data and modelling, as data are abstracted from their raw form to the higher representations used by GIS, they pass through a number of different conceptual data models and modelling via a series of transformations. Each model and each transformation process contributes to the overall uncertainty, present within the data and within analytical results. Several authors have already introduced methods for the measurement of data uncertainty as well as that of model sensitivity analysis in studies using GIS (e.g. Hwang *et al.* 1998; Gahegan and Ehlers 2000; Crosetto *et al.* 2000).

In this cross-disciplinary integration effort to master the complexity of global environmental problems, the Internet plays many important roles (Kingston *et al.* 2000). The diffusion of raw data through online geospatial databases and the existence of online mapping and management tools consist important Internet-based technological advancements that make information available to many management authorities, facilitate exchange of scientific ideas, and increase public environmental awareness. Today, the Internet is the major source of satellite imagery for marine GIS applications and hosts several GIS-based management tools specifically developed for sensitive marine areas. Bisby (2000) identified the vast amount of biodiversity-related information systems that exist today on the Internet and gave examples of collaboration among biologists and computer scientists, who have started to organise the scattered information and turn the Internet into a giant global biodiversity information system. Edwards *et al.* (2000) described one such effort as the Global Biodiversity Information Facility (GBIF), which is a framework for facilitating the digitisation of biodiversity data and for making interoperable the unknown number of biodiversity databases that are distributed around the globe. When completed, this system will be an outstanding tool for scientists, natural resource managers, and policy makers.

Several other issues accompany GIS technology, mainly issues of a technical and infrastructure nature. These issues are fuelled by the growing need for developing applications based on interaction among various hardware and software components located at different sites, owned by different vendors, and designed for widely differing application domains. They are linked to the uncontrolled expansion and evolution of data formats and GIS software as well as companies' effort to gain a greater share of the global GIS software market. The existence of a great number of different data formats and different GIS software packages puts a barrier on data exchange and integration as well as on information flow. This variety in data formats and GIS software has generated a compatibility issue among GIS developments, which limits comparison of similar analysis among scientists in an open technological environment. Interoperability enables sharing and exchange of information and processes in heterogeneous, autonomous, and distributed computing environments (Egenhofer 1999; Egenhofer *et al.* 1999). The idea aims at a cost-effective and user-friendly means to maximise the usefulness of information computing resources across multiple platforms and institutions. This is particularly important in the field of GIS since collection and editing of geospatial data is often costly and involves labour intensive and time-consuming tasks.

To achieve information interoperation for applications and end users, a wide variety of approaches has been taken, including the use of distributed object technology (Paepcke *et al.* 1996), query languages (Gingras *et al.* 1997), interface standardisation (Wegner 1996), and interface bridging (Clement *et al.* 1997). Phillips *et al.* (1999) described the concepts of data warehouses, data marts, clearinghouses, data mining, interoperability, and spatial data infrastructures, concepts that are closely related but still their differences influence their potential applications in information management and spatial data handling. In addition, efforts to interchange data among different GIS software architectures are continually in progress. For example, the development of the Hierarchical Data Format (HDF) by the National Centre for Supercomputing Applications (NCSA), which stores different data formats in one file, partially resolves the data incompatibility issue and promotes an open and free technology that facilitates scientific data archiving, exchange, and access. Interoperability presents a much greater challenge in GIS than in other fields of information science because the greater complexity of geographic information (GI) adheres to ways that acquire, represent, and operate geospatial data. The capacity of GIS to integrate and access remote data and processes transparently in an open environment is currently one of the main research and development efforts in the geographic data processing community. Efforts by the OpenGIS Consortium (OGC) have succeeded in achieving consensus within several families of applications, and some of these have now been implemented in ready to use software (Kottman 1999). Herring (1999) presented a GIS development, which is based on the concept of a comprehensive set of common software interfaces supported by geographic servers, across computing platforms. Choi *et al.* (2000) presented a component-based software system that offers GIS core functions illustrating the advantages of component-based OpenGIS. Ladstatter (2000) underlined that interoperability is far outreaching the current practice of file-based data conversion (without precluding it either) and described how the rapid development of Internet technologies influenced the OGC interoperability methods. Simple interfacing among freely available and open source software and platforms (e.g. between GRASS GIS and R statistical data analysis language in GNU/Linux) are already introduced (Bivand 2000). Model-based data transfer among GIS tools (e.g. INTERLIS) is also introduced and has led to implemented standards, which support interoperability in federated systems (Keller 1999). In addition, the use of the eXtensible Markup Language (XML), a simple, flexible, and powerful way for networked computers to exchange data and control information (W3C 1998), is introduced as a delivery method of customised RS data products to web connected clients (Aloisio *et al.* 1999) and as a real-time updating method for databases within client/server architectures (Badard and Richard 2001). However, until today, most marine GIS developments and applications are based on commercial GIS packages (e.g.

ARC/INFO, ArcView, MapInfo, IMAGINE, MGE, etc.) because of the enhanced and embedded analytical functions of such software that are essential for complex marine GIS analysis. The use of these commercial GIS software by a great number of scientists resulted in the production of many freely available computer utilities that allow data conversion among several native formats of such software.

The description of marine dynamic changes requires time series of various datasets that are often available in different data formats. Metadata, the archiving of information about a dataset, has become an issue and several different metadata standards are in use today. The purpose of metadata is to facilitate access and to guarantee appropriate application of data existing in different formats, stored in different distribution media, and are physically in different places. Since storing and retrieving geographical information has become an important part of our information society, the objective of metadata standards is to provide a clear procedure for the description of digital geospatial datasets, so users will be able to determine whether the data in a holding will be of use to them and how to access the data. Metadata standards often require the collection of several pieces of information about a dataset (mostly dataset content, spatiotemporal resolutions, associated cost, holding vendor, distribution media, etc.). Burnett *et al.* (1999) discussed the two main approaches in metadata development, that of bibliographic control (origins in library science) and data management (origins in computer science). Metadata standards exist in the form of web pages (e.g. Abreu *et al.* 2000), lists, and metadata creation tools, although efforts to develop alternative simpler, yet inclusive, globally applicable metadata standards are already introduced (Tschangho 1999).

GIS infrastructure is an important aspect of natural resources management. A proper GIS infrastructure facilitates data exchange among those public and private stakeholders whose actions significantly influence the quantity or quality of coastal and marine resources and environments. Despite the economic, social and ecological importance of our marine resources, development and management of their inherently complex dynamics is still largely pursued on a sector-by-sector basis and regulated on a jurisdictional basis. Examples of problems caused by this fragmented approach include habitat destruction, spatial conflicts, and inefficient resource use. The existence of GIS infrastructure provides an integrated platform for the horizontal (cross-sectoral) and vertical (the levels of government and non-governmental organisations) coordination of those vendors whose actions influence the marine environment. For example, Qatar is the first country to implement a comprehensive nationwide GIS and is internationally recognised as having one of the finest GIS implementations in the world having dozens of GIS applications developed and many government workers, private businesses and citizens benefited. In 1990, Qatar established a National GIS Steering Committee and The Centre for GIS aiming to implement GIS in an organised and systematic fashion and serve the GIS requirements of all government agencies. Today, many government agencies in Qatar are using GIS in their daily activities. Databases among these agencies are compatible and they are all integrated through a high-speed fibre optic network (GISnet). Another established implementation of corporate environmental GIS is that of British Columbia Environment (BC Environment), which is a part of the BC Ministry of Environment, Lands, and Parks for the protection, conservation and restoration of natural diversity, healthy and safe land, water and air, and sustainable social, economic and recreational opportunities within a naturally diverse and healthy environment. BC Environment has started to implement corporate GIS since 1993 and it has since altered the way that the Ministry handles its data focusing on a corporate approach to both geographic and attribute data organised in ORACLE databases and providing desktop access to environmental datasets across the province.

During the last decade, Marine GIS has become a well-established field of study. It includes a variety of sophisticated applications, which describe the major components of the marine environment using several brilliant approaches. In any aspect of Marine GIS, applications are widely spread among data distribution tools and online databases, mapping tools, and data integration tools for coastal, oceanographic or fisheries related tasks. The nature of georeferenced data that can be integrated and plotted to maps or inserted into visualisation systems, and the increasing use of spatially distributed approaches to environmental studies make the field of Marine GIS a promising one concerning the study of marine dynamics and management of marine resources.

The aims of this book are essentially to relate what has been done and what is now achievable from the technological perspective in oceanographic and fisheries GIS. For this purpose, the book includes many examples of excellent marine GIS developments in both fields and proposes some new developments of the use of GIS in the study of oceanic surfaces (e.g. measurement of primary productivity due to upwelling) and the use of species life history data in fisheries GIS analyses (e.g. identification of essential fish habitats and migration corridors).

## 1.2 THE ESSENTIAL GOAL OF MARINE GIS

The implementation of GIS in the marine context is multifaceted. The attributing of location to marine data and their organisation into GIS databases are the basis for a wide range of marine GIS tool developments. Indeed, a long list of goals for the development of a marine GIS may be created, the simple mapping of a parameter, the study of a oceanographic process, the explanation of species distribution patterns, to name but a few. Based on such goals for marine GIS development, marine GIS tools may be grouped into four main categories: (1) cartography tools; (2) data distribution tools; (3) monitoring tools; and (4) decision support tools. Depending on the desired output, the needs of end-users, and the longevity of the use of a GIS tool, these four categories include the main goals for a marine GIS development. There is a strong relation in the structure of these tools with a GIS database as a milestone (Figure 1.2). Each of these tools is examined in the following paragraphs.

The organisation of raw marine data in space and time on a GIS database opens up many approaches for the manipulation of the data. Marine GIS, as a cartography tool is the basic goal of GIS development. Cartography tools are very important because they can show the spatial distribution of a raw dataset as well as the spatiotemporal change of the data in a series of hardcopy maps or animated cartography. The visualisation of the spatiotemporal distribution of a dataset provides background knowledge of the nature of the dataset revealing a first picture of possible seasonal changes or features of special interest. Such tools are widely developed in the marine field with common examples in the mapping of bathymetry, the mapping of fisheries production, and the mapping of the distribution of sea surface temperature (SST) and other environmental parameters.

One step further is the enrichment of the GIS database with georeferenced data of various parameters and disciplines and preparation of GIS ready datasets for distribution. GIS data distribution tools may be developed in the form of marine GIS atlases and other archiving GIS databases, which may then be distributed by means of CDROM and web interfaces via the Internet (online data servers). The great importance of such applications is in the grouping of an ever-increasing amount of different georeferenced data. Besides GIS data, these tools often provide lists of metadata (information about the nature of included data) simplifying the data search process and ensuring that users can find the exact GIS ready data needed for their tasks. Data distribution GIS tools can be seen as value adding tools because they provide raw data in GIS ready format, thus enhancing the use of the raw data, especially of satellite data.

From this point, marine GIS may be developed as monitoring tools. Time series of GIS datasets may be analysed for the monitoring of environmental processes establishing the seasonality of certain oceanographic phenomena such as the start and the end of a cyclonic gyre or an upwelling event. In addition, such tools can be used for the monitoring of marine species population dynamics, their stocks, and their fisheries production levels (catch and landings). Monitoring of such parameters is important for the knowledge of the current state of marine resources, the seasonal, annual, and interannual cycles of oceanographic phenomena, and how they relate to marine species life cycles. Extensive database structures and high levels of data integration characterise such GIS developments. The aim of these tools is to turn raw data into meaningful information. Extensive data integration and spatial analysis give results for the description of marine dynamics revealing seasonal relations among biotic and abiotic elements of the marine environment.

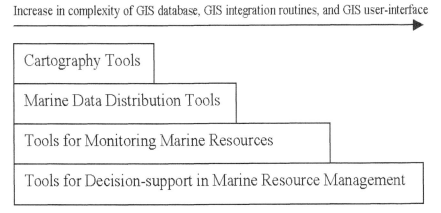

**Figure 1.2.** The four main purposes of developing a marine GIS tool. The common part of such tools is a GIS database, however, the complexity of tools' design increases according to the specific development purpose.

All previous types of GIS tools are included in a marine GIS decision support tool. Such tools are extremely valuable for the development of marine resource management scenarios that are based on information about the dynamics of marine resources. These tools are specialised marine GIS developments often used for studies in small areas or of a particular species or family. They provide in-depth analytical results for species population dynamics, their life cycles in relation to the marine environment, their past and current stock geodistribution levels and fisheries production status. In addition, these tools have two main specialties: First, they reveal sensitive areas during a species life cycle, such as a species seasonal spawning grounds as well as overfished areas and remote, underfished grounds of species richness. Second, by integration of the spatial extent of current management policies, these tools reveal possible conflicts between currently applied management schemes and species population dynamics. The importance of these two main specialties is in providing information for the adjustment of current management, creation of a new information-based type of management, and the development of forecasting statistics.

Based on this approach, it could be said that the essential goal of a marine GIS is that of generating information-based management proposals with inherent support from GIS cartography, data distribution, and monitoring tools. Information-based management of marine resources as well as the use of new technologies (such as GIS and RS) towards this goal has been constantly emphasised during the last two decades. For example, both the United Nations Convention on the Law of the Sea (UNCLOS) and Agenda 21, from the United Nations Conference on Environment and Development held at Rio de Janeiro in 1992, call for information-based approaches to management. Nations have begun to respond. Canada's new Oceans Act, for example, passed in January 1997, includes specific provisions for information-based management. The same applies to the philosophy of several framework programmes, such as the European Commission's Fifth Framework Programme and the Short and Medium Term Priority Environmental Action Programme or the US National Science Foundation various ocean programmes. Integrated spatiotemporal output from marine GIS applications becomes of vital importance to decision-making.

## 1.3 SPATIAL THINKING AND GIS ANALYSIS IN THE MARINE CONTEXT

An understanding of the dynamics of marine phenomena and marine living resources as well as the explanation of the relationships among the marine environment and species populations are essential in any human attempt to manage marine natural resources. The development of spatial thinking becomes very important in various levels throughout the study of processes in the marine environment through GIS, from data sampling to GIS development and decision-making. The material presented here is based on the work of several authors who studied spatial cognition in geographic environments (Slater 1982; Golledge and Stimson 1987; Blades 1991; Nyerges 1994; Golledge 1995; Nyerges and Golledge 1997; Lloyd 1997). Spatial thinking and GIS bring a totally new approach to the management of natural resources, devaluing 'experimental' management to information-based management.

Maybe the understanding of the importance of geospatial (geolocated) data is the first step for the development of spatial thinking. The nature of any natural or socio-economic activity with a spatial dimension cannot be properly understood without reference to its spatial dimensions. The two essential parts of spatial data (location and associated attributes) are used to place information on database management systems that contain location under a typical locational reference system (e.g. latitude and longitude, area or distance specific projections) linked to attributes stored under a storage format (e.g. tables). Thus, phenomena that contain a spatial component may be studied through GIS and shown on maps. There are six basic concepts that are inherently spatial and are used by geoscientists in studying spatial phenomena (Nyerges and Golledge 1997). These spatial concepts are location, distribution, region, association, movement and diffusion.

The most basic spatial concept is that of location. For example, the location of a meteorological station will give a spatial meaning to the associated dataset. Also, the first question, for example, that an oceanographer studying an upwelling event will typically ask is 'where does it occur'? Distributions may be thought of as sets of individual locations of one or more datasets describing a part or the whole of an area. A region is an area that is distinguished from other areas by one or more characteristics. By creating a region, a scientist is able to generalise and simplify. A region, for example, is an area where SST is generally lower than in the surrounding area. If we have two different spatial distributions that appear to be similar, we have a spatial association. For example, the phenomenon that SST is low in the same area where surface chlorophyll concentration (CHL) is high attributes a spatial association between the two phenomena. Of course, this does not prove a cause and effect relationship, but may give a reason to attempt to understand why the association exists. Movement from one area to another is also something that is inherently spatial. For example, the migration of fish populations is one form of movement of interest to fishery biologists. Finally, diffusion is the process by which something spreads. The phenomenon of cold patches of SST that are diffused along a coastline is of interest in physical and biological oceanographers. Based on these six concepts, a geoscientist may integrate the associated datasets and explain, for example, how the distribution of wind data from several locations of meteorological stations affects a region where a spatial association between SST and CHL does exist and how movements of fish populations correspond to this process, and finally, how this process is diffused in space and time.

One of the best ways of comprehending and developing spatial thinking is to learn to ask geographic questions. Such questions encourage thinking and learning by posing a problem that requires an answer. In turn, answers involve the creative integration and manipulation of GI by connecting facts or constructing scenarios for the final answering of a question. The set of questions that Slater (1982) suggests should be in every inventory with a geographic content along with examples of marine spatial questions are presented in Table 1.2.

**Table 1.2.** Suggested spatial questions integrated in every geographic inventory (Slater 1982) and associated examples of marine spatial questions.

| SLATER (1982) SUGGESTED SPATIAL QUESTIONS | EXAMPLES OF MARINE SPATIAL QUESTIONS |
|---|---|
| Where is it? | Where is the location of a meteorological station and the associated data? |
| Where does it occur? | Where is the location of the centre of a gyre or an upwelling? |
| What is there? | What is the topography of the area where an upwelling occurs? |
| Why is it there? | Why does upwelling consistently occur in a particular area? |
| Why is it not elsewhere? | Why upwelling does not occur in all coastal areas? |
| What could be there? | What are the wind patterns and bathymetry of an upwelling area? |
| Could it be elsewhere? | Are there areas with similar conditions where upwelling may occur? |
| How much is there at that location? | How many trawl vessels fish in a particular area? How much fish weight is landed in a fish market? |
| Why is it there rather than anywhere else? | Why do trawlers consistently fish in a particular area and not elsewhere? |
| How far does it extend already? | What is the surface area of an upwelling? How far does a temperature front extent? |
| Why does it take a particular form or structure that it has? | Why do gyres appear as cyclonic or anticyclonic formations? Why do fronts form long stripes on sea surface? |
| Is there regularity in its distribution? | Is there seasonality in a gyre's formation? Is there a particular area where a specific marine species is consistently caught? |
| What is the nature of that regularity? | What are the characteristics of a gyre's seasonality (e.g. when the gyre starts and ends, how strong or weak it is)? |
| Why should the spatial distributional pattern exhibit regularity? | What are the environmental characteristics that cause seasonality in a gyre formation or upwelling event? |
| Where is it in relation to others of the same kind? | Where is the location of a cyclonic gyre in relation to a nearby anticyclonic gyre? |

| | |
|---|---|
| What kind of distribution does it make? | What is the distribution of sea surface temperature, chlorophyll, and salinity before, during, and after an upwelling event? |
| Is it found throughout the world? | Does upwelling occur only in certain coastal areas? Is trawl fishery for a particular species localised or is it found throughout the world? |
| Where are its limits? | Where are the spatial and temporal limits of trawling activity in a particular area? Where is the maximum depth of occurrence for a particular marine species? |
| What is the nature of those limits? | Do bathymetry and distance-from-coast limit trawling activity? |
| Why do those limits constrain its distribution? | Why is geodistribution of a particular species limited by bathymetry? |
| What else is there spatially associated with that phenomenon? | Is a front associated with an upwelling event? |
| Do these things usually occur together in the same places? | Do temperature and chlorophyll anomalies occur in the same places? Do fish aggregate in upwelling areas? |
| Why should they be spatially associated? | Why temperature anomaly is an indication of a front? Why upwelling is an indication of fish aggregation? |
| Is it linked to other things? | Does chlorophyll concentration increase when temperature decreases? |
| Has it always been there? | Is there a persistent gyre in a particular area? |
| When did it first emerge or become obvious? | When did a particular upwelling area start having effect on productivity? |
| How has it changed spatially (through time)? | How have productivity levels changed in a particular area? |
| What factors have influenced its spread? | Do wind duration and direction and bathymetry distribution influence the spread of an upwelling? |
| Why has it spread or diffused in this particular way? | Why does an upwelling spread 1 nautical mile or 10 n.m. from a coast? Why does the epicenter of a gyre move eastward? |
| What geographic factors have constrained its spread? | Do wind direction and force constrain the development of an upwelling event? Why are particular species found 200 n.m. from their spawning grounds? |

Nyerges (1994) suggests that such geographic questions may be categorised into five groups: (1) those dealing with location and extent; (2) distribution and pattern or shape; (3) spatial association; (4) spatial interaction; and (5) spatial change. To both ask and answer geographic questions, one should use the template of concepts used in geography, e.g. location, distribution, pattern, shape, association (Golledge 1995) and an outline of the processes involved in thinking geographically, e.g. observing, defining, interpolating, spatially associating (Nyerges 1994). Together the template and the process assist not only in handling specific questions but also in linking questions that may not otherwise appear to be linked.

GIS can help form, generate, and define geographic questions as well as help solve them by enabling representations of data to be displayed and visualised. In this way, GIS helps with identification and definition (e.g. generating 'what' and 'where' questions) as well as in solving them by using various display modes. In turn, questions of association can be illustrated with overlay procedures and questions of change can be generated from sequential 'snapshots' of locations, patterns and distributions. A variety of analytical functions help solve 'why' questions, and a selection of methods can be used to examine questions of process and interrelation. Thus, GIS can 'phrase' questions in different formats (graphic, pictorial, mathematical) and use certain methods for answering certain problems (Golledge and Stimson 1987).

Two other important steps in developing spatial thinking are the understanding of various GIS data models (Blades 1991; Lloyd 1997) as well as the appreciation of spatial analysis in GIS (Cances *et al.* 2000). The definition of questions, the processing of the associated data, and the reaching of answers are all linked to the structure of a GIS database. Such databases store georeferenced data represented in vector and raster formats. Vector formats include representation of data as points, lines, polygons, and regions while raster formats represent data in picture elements (pixels) or volume elements (voxels). The main types of marine data formats that are stored in GIS databases are listed and shown in Table 1.3 and Figure 1.3, respectively. From that point, GIS can manipulate these data in a variety of ways, which include conversion among data formats (e.g. images converted to grids), creation of new data formats (e.g. TIN, Triangular Irregular Networks), preparation of data for analysis (e.g. creation of grid stacks) and finally, integration analysis of vector and raster datasets.

Spatial analysis in GIS refers to a large number of modelling operations in one or more datasets via a sequence of elementary actions, which are important for decision support in management. Spatial analysis in most GIS packages works on both vector and raster data through construction of data topology (e.g. georeferenced grids, connected arcs, closed polygons). Spatial analysis tools in GIS include a variety of techniques, such as classification and aggregation, proximity analysis, adjacency analysis, connectivity analysis, optimum path analysis, statistical analysis, interpolation and outlining as well as various data integrations.

Classification involves reassigning a data value to a descriptive attribute of a polygon according to the values taken by other attributes. Classification can be followed by aggregation, which involves grouping two or more adjacent reclassified units by dissolving boundaries between polygons and then reconstructing new data topology. For example, an SST image can be classified based on certain data ranges and then aggregated to areas of low, medium and high temperature values (e.g. identification of upwelling areas). Proximity analysis involves determining several spatial features (points, lines, polygons) located within a maximum distance from a given spatial feature. Proximity analysis introduces the concept of a 'buffer zone', which can be of a fixed or variable size. For example, proximity analysis may be used to locate fishing ports that are within 100 miles from a given fishing area (e.g. spatial relation between catches and landings). Adjacency analysis involves reassigning to a given data value a new value, which depends on that of neighbouring data values. In raster data, such as satellite images and aerial photographs, adjacency analysis might employ a filter, such as a summation, mean or gradient filter. For example, adjacency analysis may be used for the computation of slope (e.g. the rate of change in data values) in a CHL concentration image (e.g. identification of CHL fronts). Connectivity analysis consists in determining the boundaries of an area by starting from a certain data value and moving in every direction in order to locate the points verifying a given data value (e.g. threshold value). For example, connectivity analysis may be used for the identification of areas where SST is between 20 and 23 °C.

**Table 1.3.** Various GIS data formats and example marine-related datasets.

| DATA FORMAT | EXAMPLE DATASET |
| --- | --- |
| Vector-point | Wind measurements (direction and force) and any other point measurements for sediment types, depth, etc. |
| Vector-line | Coastline and bathymetry |
| Vector-polygon | Fisheries data sampling schemes (e.g. statistical rectangles) |
| Vector-region | Spatial extent of fishing gears activity and spatial extent of fishery laws |
| Raster-pixel | Satellite imagery and aerial photography |
| Raster-voxel | CTD measurements and sonar/hydroacoustic data |

Optimum path analysis consists in determining the optimum path between two points or areas considering distance, cost, time and other factors. For example, when combining fish migration habits and environmental data, this type of analysis applies well to seasonal fish migrations. Impedance, which is defined by characteristics such as certain data values, is assigned to each data value revealing, e.g. the difficulty of a species to pass through certain data values of SST (according to species preferred environmental conditions). Statistical analysis supplies basic statistics on the descriptive data such as mean, standard deviation, minimum, maximum and median values and histograms of data distribution. More sophisticated analyses such as regression, classification and Principal Component

Analysis (PCA) provide valuable information, for example in relations between fisheries populations and oceanography. These calculations are common in image processing software packages and are widely used in GIS. However, current GIS packages often do not include extensive statistical tools but do provide data exchange with such tools. Interpolation techniques provide an estimate of a value at a point where the value is unknown, within a region covered by a number of known values of sampled points. The choice among the many interpolation methods depends on the spatial model, which is best fitted to the sampled values. Two commonly used interpolation methods are local interpolation (spline, weighted mobile mean), and kriging. These methods are based on the hypothesis whereby two points, which are close to each other are more likely to have similar values for a given property than two distant points are. They take into account the fact that space is not necessarily isotropic (e.g. the bottom sediment values from a point sonar survey are more likely to vary through the whole of a study area than between two closely together and parallel transect lines).

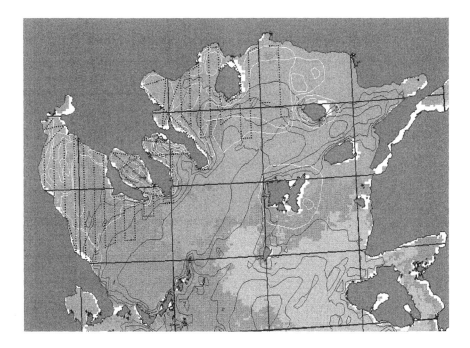

**Figure 1.3.** Representation of various marine data formats in GIS. Points: survey data (in black); Lines: bathymetry (in black); Polygons: fisheries statistical rectangles (in black); Regions: fishery tools activity areas (in white); Pixels: background image of SST distribution. Land is shown in dark grey.

GIS data integration, the central point of GIS technique, is what turns raw datasets to meaningful information. The GIS ability to spatially integrate two or more datasets (raster and vector) is highly suitable for marine GIS analysis. Processes and relationships in the marine environment are characterised by continuing change in the spatial and temporal distribution of several environmental parameters, e.g. wind pattern, currents, temperature, CHL and salinity. Integrated analysis of such parameters through GIS reveals the spatial extent of several oceanographic processes, e.g. upwelling, gyres and fronts. For example, combined classification of temperature and CHL satellite imagery reveals areas of particular oceanographic interest. Addition of spatially referenced fisheries catch data reveals spatial and temporal relations between species populations and oceanography. Further addition of species life history data reveals areas of particular interest in fisheries management. In GIS, the extraction of this information from raw data becomes possible through various data selection and integration techniques, which include combinations of selective geometric intersections, data erasing and data union. These techniques are thoroughly explained with various examples in this book.

Consequently, spatial thinking and GIS analysis in the marine context are very important for understanding the nature of the dynamics of marine processes and how these affect the dynamics and behaviour of species populations according to their life history characteristics. During the process of marine GIS development, spatial thinking and analysis often require multiple inputs. For example, the explanation of why a mobile cephalopod species (*Loligo vulgaris*) is found 200 km offshore at certain times of their life cycles, requires the input of species life history data into their migration habits (fisheries biologists), wind and current patterns, and the identification of food supply upwelling events in the region at that specific time (physical and biological oceanographers) as well as expert GIS developers for the integration of this knowledge. From the GIS point of view, the ultimate aim is to join all required knowledge and develop a model of the marine environment in order to understand what and where things are and how and why they are where they are.

## 1.4 CONCEPTUAL MODEL OF A MARINE GIS

Marine GIS development, as a multidisciplinary process, requires the involvement of a scientific team of oceanographers, marine biologists, and GIS experts. During this teamwork procedure, several issues become important. The first task of a marine GIS developer is to cooperate with other marine scientists for the identification and definition of the spatial problem and the creation of a list of spatiotemporal questions that will be examined through the marine GIS tool. The nature of these questions will greatly affect the whole design of the marine GIS tool because such tools contain specialised GIS tasks for answering specific questions. Thus, the list of questions will define the specific datasets that will be gathered and used by GIS for reaching answers, the design of the GIS database and the development of the GIS data manipulation routines that will be specifically created for each question. The nature of the questions as well as the system's end-user requirements will define the design of the tool's user interface. Interface design and development are important to the effective and friendly use of the whole system.

Consequently, a marine GIS is characterised by three components: (1) the spatiotemporal multidisciplinary database structure; (2) the data manipulation routines; and (3) the system user interface. Each of these components is equally important to the success of a marine GIS because they provide a finished and usable product to the hands of researchers and policy officials. Because of its multidisciplinarily nature, the design of a marine GIS database should provide automated data input and georeference as well as multiple data selections so that data manipulation routines process only the needed fraction of the GIS database for a specific task. Keeping things automated and simple at the same time is often a difficult task for GIS developers, resulting however in a usable and effective tool. In addition, the data manipulation routines should be used in a 'step-by-step' basis, so that intermediate outputs can be checked before used as input in the next data integration. This will reduce analysis error propagation to the final output. Finally, all marine GIS routines should be interfaced, so that users can communicate with the GIS database in a user-friendly and selective way.

The dynamics of the marine environment are linked to the seasonality of oceanographic processes and the biology of species populations in a 3D space. Thus, a marine GIS tool is called upon to answer spatiotemporal questions about these dynamics and examine their relations. In a computerised setting, marine GIS is called upon to absorb, store, analyse data and present results about the relations of marine processes and populations. For this purpose, the spatiotemporal multidisciplinary database structure of a marine GIS must store time series of marine environmental parameters and species population statistical, biological, and genetic data.

An initial analysis step is the examination of oceanographic processes. The identification and measurement of such processes (upwelling, fronts) are vital for understanding the seasonality of marine processes and creating a clear picture of what type of process it is, when and where it occurs, how strong or weak it is, and how long it lasts. This analytical step often requires the manipulation of image data, such as image georeference, rectification, projection, and application of edge detection algorithms as well as processing of point data, such as the application of spatial interpolation methods. At this stage, the integration of multiple datasets (e.g. temperature, CHL, salinity) is essential to the study of oceanographic processes. A following analysis step is the examination of species population dynamics. A good start for this analytical step is the use of species life history data. These data are provided by biological and genetic research on species life cycles and are continually updated when recent new information becomes available. Species life history data include important information on species biology and ecology. Information on species spawning preferences, optimum living conditions, occurrence maximum depth, migration habits, etc. is highly appropriate for marine GIS analysis. These data can give information on species occurrence areas and when combined with other datasets (e.g. major fishing activity areas) reveal species suitable habitats, spawning grounds, and migration corridors. Following, the integration of findings from the previous analytical steps reveals the spatiotemporal relation of species distribution to oceanography and maps the seasonal change of species concentration areas according to varying oceanographic processes. As the level of generating management scenarios through GIS is approached, the spatial extent of current marine and fisheries policies should be further integrated in a marine GIS development (Fowler *et al.* 1999; Guillaumont and Durand 1999).

Based on the suggested conceptual model of marine GIS development, such a system should incorporate a variety of raster and vector datasets. These datasets are listed in Table 1.4. In addition, the manipulation of these datasets should be such that the developed marine GIS analytical routines include constraints from species life history data. In this way, the modelled GIS output will be based on known species life cycles. During the marine GIS development process, the utilisation of the Internet should be taken into account, especially for data downloading and GIS result dissemination. Today, several data providers disseminate their data holdings through Internet data servers. A variety of datasets can easily be located and downloaded through the Internet. The Internet is also a suitable medium for the dissemination of GIS output. This is particularly important in communicating GIS output through different management authorities that share jurisdiction in coastal and fisheries areas. A diagram of the complete architecture of the proposed conceptual model of marine GIS development is presented in Figure 1.4.

**Table 1.4.** Standard vector and raster datasets, which are often easily accessible and should be included in any marine fisheries GIS database.

| DATASETS | COMMON GIS FORMATS |
| --- | --- |
| Coastline | Line or Polygon |
| Bathymetry | Line or Polygon |
| Sea Surface Temperature | Raster |
| Sea Surface Chlorophyll | Raster |
| Sea Surface Salinity | Raster |
| Sea Surface Wind/Currents | Raster |
| Location of Fish Markets | Point |
| Commercial Landings | Point |
| Commercial Catch | Polygon |
| Statistical Sampling Areas | Polygon |
| Extent of Fishing Fleet Activity | Region or Polygon |
| Extent of Maritime Policies | Region or Polygon |
| Species Life History Data | ASCII Table (source of constraint parameters in GIS analysis routines) |

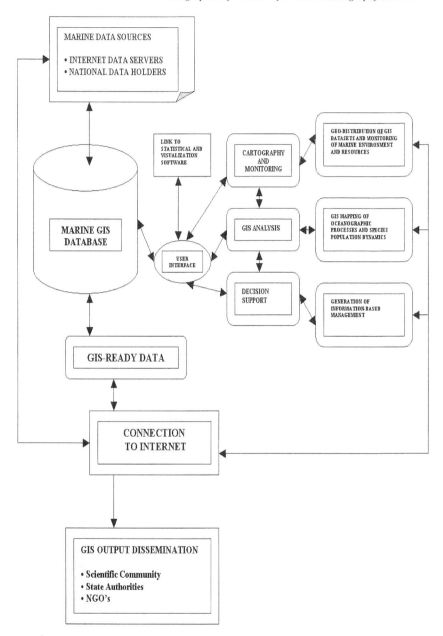

**Figure 1.4.** Chart flow diagram of a complete architecture of marine GIS development for oceanography and fisheries related tasks.

Suffice it to say that a marine GIS development for the study of oceanographic processes and fisheries dynamics should incorporate the following tasks:

- display marine (oceanographic and fisheries) data, including station plots, model grids, radar and satellite images (e.g. surveyed, modelled and remotely sensed data);
- use intrinsic GIS functionality to support traditional analysis functions, including contouring, cross-sections, superposition (overlay), and integration;
- incorporate user supplied algorithms and software through calls to other subroutines;
- incorporate links to GIS external software (statistical and visualisation);
- support an interactive graphic user interface for friendly use;
- support static representation as well as animation of results; and
- enable data downloading and result dissemination through the Internet.

GIS is a complex tool that requires careful planning and design to be successfully implemented. Choices in hardware, software and data must be based on evaluation of project objectives, analytical requirements, data availability and data development considerations. Datasets must be evaluated and documented with metadata. The various steps in data transformations and GIS data analysis must also be documented.

## 1.5 GIS AND SCIENTIFIC VISUALISATION SYSTEMS

Spatial data and products of GIS have become a powerful resource that is being utilised at all levels within our society. Effective environmental decision-making for regulatory purposes is a real international problem faced on a daily basis on the global, national, state and local levels. The integration between the analytical functionality of GIS and image processing with the interactive display and analysis of scientific visualisation can provide a valuable tool for resource management, monitoring and modelling. Marine applications have three notable characteristics: (1) the amount of data is increasing dramatically, posing performance and analysis problems; (2) the data are complex and heterogeneous representing multiple formats, scales, resolutions, and dimensions; and (3) a wide range of personnel must interact with the data as solutions to marine spatial problems increasingly involve interdisciplinary teams. The current software infrastructure for such applications does not adequately address these three characteristics, making enhancements and extensions to current products mandatory. One extension, which is gathering force and interest, is the integration of GIS and Scientific Visualisation Systems (SVS).

While the integration of SVS and GIS methodologies brings with it great opportunities, it also presents significant problems in effectively communicating information, particularly to non-GIS professionals, such as policy officials. Current GIS technology provides a variety of mechanisms for 2D data integration and visual communication. However, the human visual system inherently is capable of using a variety of perceptual cues to extract information within a full 3D environment. Unfortunately, the availability of hardware and software to support this kind of powerful visualisation has not existed until very recently. By viewing GIS data in new unconventional ways, new modes of analysis and exploration are enabled. This can provide a capability to enhance the use of existing GIS data as well as provide a motivation for extensions to GIS database models themselves.

The SVS and GIS methodologies are frequently used to examine marine sciences data. Both disciplines developed and have often been implemented in parallel to each other, however being slow to exchange ideas. Where GIS provide strong support for analytical functions, SVS provide the capabilities to visually interact with the data using a variety of sophisticated techniques. These systems can support multidimensional data, providing the infrastructure to generate 2D and 3D spatial views, animations of time series, and graphical or statistical plots. However, the lack of analytical operations such as data integration in three dimensions or queries across time series still limits the capabilities of these tools. Bringing the analytical strengths of GIS together with the visualisation strengths of SVS would provide a more robust set of software tools for marine applications.

Three levels of GIS and SVS integration have been identified: (1) rudimentary; (2) operational; and (3) functional. The rudimentary approach uses the minimum amount of data sharing and exchange between the two technologies. The operational level attempts to provide consistency of the data while removing redundancies between the two technologies. The functional form attempts to provide transparent communication between these software environments. These levels of integration offer many advantages. One is the ability to generate visualisations much faster since data do not have to be converted from a non-native format. This is especially important for animations where a smooth transition between time steps yields a more effective visualisation. Another advantage of SVS and GIS integration is the link to the GIS database. With this link, spatial or logical queries can be made to the GIS allowing for subsets or different spatial attributes to be visualised easier. A third advantage is the link to other system functionalities, such as statistical processing that could be used to perform analysis on attributes and attribute combinations. Visualising the results of these analyses could be the most effective way to understand complex attribute relationships. A fourth advantage is that attributes and spatial resolutions need not be compromised. With a direct link to the GIS database, no filtering is necessary and visualisation can take place using all data available. And this is what visualisation excels at, making sense out of extremely large amounts of information. Thus, effective data visualisation requires close coordination with GIS database management systems. GIS databases are of particular interest because with them, data can be stored in a way that makes their subsequent use not only possible but also efficient. They offer the capability to search for data based on content rather than solely by name, something that is crucial to enhancing understanding. This act of browsing is an important interactive visualisation function and as such, data visualisation systems can be thought of as portals into GIS databases and functionality.

The importance of GIS and SVS integration in decision support is already recognised (e.g. Mason *et al.* 1994; Kraak *et al.* 1995). Current efforts in integration of GIS and SVS are multifaceted. Shyue and Tsai (1996) discussed the functional demands of 2D and 3D marine GIS and summarised the published 3D GIS data models. They used Intergraph MGE Voxel Analyst (MGVA) software to show the 3D distribution of sea temperature. Bernard *et al.* (1998) presented an interoperable object-oriented GIS framework for atmospheric modelling (AtmoGIS), which is based on a spatiotemporal database management system, a mesoscale model and an environment for scientific visualisation. Breunig *et al.* (1999) proposed a component-based GIS, which is coupled with an object-oriented database management system where 3D modelling tools have direct access to a 3D database via the Common Object Request Broker Architecture (CORBA). Also, Breunig (1999) introduced GeoToolKit, an open 3D database kernel system implemented by using the object-oriented database management system ObjectStore1. Huang and Lin (1999) designed GeoVR, a tool that uses ArcView Internet Map Server and ArcView 3D Analyst under a client/server architecture to enable the interactive creation of a 3D scene using a virtual reality modelling language (VRML) model from 2D spatial data by integrating Internet GIS and HTML (HyperText Markup Language) programming. Lees (2000) created Geotouch, a UNIX-based freely available software (also ported to Linux) that includes methods for cutting cross-sections at arbitrary angles, spinning objects in three space and animating time series of punctual data. Morris *et al.* (2000) identified the limiting factor in nearly all conventional GIS that time and depth are handled as attributes to a geographical object (e.g. point, line, area). Under LOIS (Land/Ocean Interaction Study), a UK research project investigating forms and processes in the coastal zone, they created the SpatioTemporal Environment Mapper (STEM), a GIS system that handles time or depth visualisation of an entity in addition to mapping the entity horizontally. This treats time or depth as a dimension rather than an attribute, which is a prerequisite to effective multidimensional visualisation and analysis. STEM is a GIS data viewer fronting a database containing the highlights of LOIS.

Visualisation has become an integral part in many applications of GIS. Due to the rapid development of computer graphics, visualisation and animation techniques, general purpose GIS can no longer satisfy the multitude of visualisation demands. Dollner and Hinrichs (2000) proposed that GIS have to utilise independent visualisation toolkits and showed how GIS can provide visualisation and animation features for geoobjects by embedding visualisation systems using object-oriented techniques. Su (2000) integrated ARC/INFO, ArcView and Vis5D for the incorporation of the various data acquisition and database systems of the Monterey Bay Aquarium Research Institute. Vis5D is a freely available visualisation system, which is widely used by scientists to visualise the output of their numerical simulations of the Earth's atmosphere and oceans (Hibbard 1998).

SVS and GIS are not necessarily mutually exclusive. The difference in technology may be more on emphasis than substance. GIS functions include data formatting, management, analysis, and display. Scientific visualisation becomes an integral part of GIS, spatial analysis and decision-making. For the long-term, integration needs to be incorporated in all aspects of spatial and temporal analysis. Statistics, GIS, modelling, remote and *in situ* monitoring, visualisation, and other tools that are now discrete functions of specialised communities need to be seamless. The highly complex and interrelated marine problems require scientific work in a networked, collaborative, interdisciplinary community, with common tools to apply the scientific method. Integration of GIS and SVS software seems to be the initial step in this effort.

## 1.6 SOFTWARE ISSUES AND PACKAGES FOR MARINE GIS DEVELOPMENT

There is a great number of GIS packages on the market today, which include commercial products from well-established software enterprises as well as some free products available from new companies willing to attract users and conquer a share in the world's GIS software market. Several products are already widely used for the development of marine GIS applications (e.g. ARC/INFO, ArcView, MapInfo, Imagine, GRASS, MGE, IDRISI, ILWIS) and through the years, they established a main core of marine GIS developments that are based on these products. Of course, this does not limit the introduction of new GIS software on the market, since the GI sector is continually growing with fast GIS solutions needed and with software learning curves and friendliness of use playing important roles to the software selection process. Specifically for the development of marine GIS software packages, the market is widely open because as of today, there is not a GIS software, which offers in one package a 3D database, 3D data integration and 3D visualisation capabilities. On the other hand, the well-established software enterprises are trying to confront with GIS interoperability issues aiming to produce products that satisfy the exchange of applications that are based on different software solutions. For example, several of these enterprises are trying to include into their products OpenGIS Consortium's interoperability specifications, which are listed at: http://www.opengis.org/techno/conformance.htm.

Throughout the GIS software selection process, several issues become important on all application levels, including single user desktop, departmental, local authority and governmental levels. Software solutions should be examined for their learning curve, ease of use, hardware platforms, network capabilities, the amount and formats of data that they can process and their connection abilities with other GIS external packages (e.g. statistical or visualisation software). For example, today most high-level GIS networks use powerful GIS packages as main data servers and a series of compatible satellite GIS software as peripheral tools for interaction with the main data servers. A practise of 'what others already use' is common to the selection process, mainly due to compatibility issues in similar disciplines.

Designs of new GIS installations for high-level networks should look in the future. The selected GIS packages should be capable to interchange data and output with other widely used packages while confronting interoperability specifications on database design, metadata documentation, data analysis, and Internet dissemination. In parallel, established and new GIS software enterprises should focus on the development of slightly different versions of their products based on the target discipline. This becomes essential in disciplines like Oceanography and Fisheries, which need extensive 3D GIS operations for the modelling of real-world dynamics.

Following, a small reference list of commercial and freely available 'lighter' GIS packages are presented, which all are worth trying for their specific level of development and purpose of use. AGIS (http://www.agismap.com/) is a simple mapping and GIS software package for Windows based applications. Data are inserted in the form of ASCII text files and maps can be animated and projected to several map projections. A built in scripting language allows the creation of map animations and link to other applications such as database queries. CARIS GIS (http://www.caris.com/) provides a marine GIS for retrieval of geographic and textual information from various databases, based on selection criteria, such as source, scale, area, depth, etc. ERDAS Imagine (http://www.erdas.com) is an image processing and GIS package, which includes many raster and vector analytical operations as well as an extensive modelling module. ESRI products (http://www.esri.com) include a variety of widely used GIS packages with build in extensive analytical tools, conversion among different data formats based on specific analysis needs and customisation languages that allow integration of both raster and vector datasets. GRASS GIS (Geographic Resources Analysis Support System, http://www.baylor.edu/grass/) is an open source GIS with raster, topological vector, image processing, and graphics production functionality that operates on various platforms. Data are inserted in the form of 2D and 3D raster files including a variety of formats. GRASS is mainly used as spatial analysis, map generation, and data visualisation tool. MapInfo products include a variety of widely used GIS module applications with integration capabilities of both raster and vector datasets (http://www.mapinfo.com/). PCRaster is a raster-based GIS where data cells can store more than one attribute (http://www.geog.uu.nl/pcraster/tekst.html). GIS operations can be performed between attributes on one cell location or between cells. SADA (Spatial Analysis and Decision Assistance, http://www.sis.utk.edu/cis/sada/) is a freeware that incorporates tools from environmental assessment fields into an effective problem-solving environment. The software includes modules for visualisation, geospatial and statistical analysis, cost/benefit analysis, sampling design, and decision analysis. SPRING (http://sputnik.dpi.inpe.br/spring/english/home.html) is a GIS and RS image processing system. It features an object-oriented data model under which both raster and vector data representations can be integrated. ProGIS (http://www.progis.com/) among other products, offers WinGIS, a mapping software tool with capabilities such as layer grouping and object selection that also accepts several common data formats. IDRISI (http://www.clarklabs.org/) offers exceptional raster analytical functionality for both GIS and RS needs and includes tools for database query, spatial modelling, image enhancement and classification. ILWIS (http://www.itc.nl/ilwis/ilwis.html) is also an RS and GIS software, which integrates image, vector and thematic data in one powerful package that delivers a wide range of capabilities, such as data import and export, digitising, editing, analysis and display of data and production of quality maps. Last but not least, TNT products (http://www.microimages.com/) include a variety of applications for both GIS and RS with many capabilities, such as vector and raster operations, attribute and database operations including a customisation language (SML, Spatial Manipulation Language). Most of the enterprises that develop these products offer fully functional trial versions or reduced functionality freeware products, often with extensive documentation.

## 1.7 SUMMARY

Marine GIS applications share only about one-third of GIS history. They created a new application theme in the field of GIS posing several challenges in the domain of Geographical Information Science. These new challenges originate from the fact that a computerised application is called upon to model the dynamics of marine environment and provide meaningful explanations about these dynamics. The opposing strength of marine GIS is that it fully develops the central point of GIS technology, that of georeferenced data integration, to explain the dynamic relation between marine processes and species populations.

Marine GIS is a highly adaptive process. In people level, it requires the cooperation of scientists from many different disciplines. Several marine scientists are needed for the development of a completed marine GIS application (fisheries and marine biologists, biological and physical oceanographers, GIS developers and analysts, etc.). In technology level, marine GIS require the manipulation of many different data formats (raster and vector) as well as the development of innovated marine GIS integration routines in order to produce output with high accuracy.

Marine GIS is also a self-enhancing process. It develops spatial thinking through the visualisation of various georeferenced raw data. By this process, marine GIS links questions that may not otherwise appear to be linked. In turn, the development of spatial thinking affects the whole design of a marine GIS with a main impact in its data integration routines as well as the final outputs. Depending on the goal of marine GIS development, marine GIS tools contain specific analytical tasks that require the creation of a list of specific spatiotemporal questions. The nature of these questions characterises a marine GIS tool as a cartography, data distribution, monitoring, and decision support tool. The integration of GIS methodologies with other disciplines (e.g. scientific visualisation, statistics and spatial analysis, modelling, RS) is well recognised as an important part of the decision support process for the study of local and global marine environmental problems and management of marine resources.

Commensurate with accelerating advances in RS, satellite telemetry, SVS and spatial statistics, the primary objective of the Marine GIS domain is to evaluate and apply these state of the art tools for developing or improving the methodologies used in marine research. The need for cost-effective techniques to systematically acquire environmental data for large marine areas, and monitored data for a great number of marine species, crosses agency, programme and issue boundaries. Application of these evolving and advancing technologies provides innovative methods to address complex natural resource issues and provide better information towards making well-informed management decisions.

## 1.8 REFERENCES

Abreu, J., Scholten, H., van den Eijnden, B. and Gehrels, B. (2000). Combining spatial metadata search engines with webmapping. *Geo Informations Systeme*, **13(5)**, 8–11.

Aloisio, G., Mililllo, G. and Williams, R.D. (1999). An XML architecture for high performance web based analysis of remote sensing archives. *Future Generation Computer Systems*, **16**, 91–100.

Asrar, G.R. (1997). Global change research and Geographic Information Systems requirements. In J.L. Star, J.E. Estes and K.C. McGwire, eds. *Integration of Geographic Information Systems and Remote Sensing*, pp. 158–175. Cambridge University Press, Cambridge.

Atkinson, P.M. and Tate, N.M. (1999). *Advances in remote sensing and GIS analysis*, p. 273. John Wiley and Sons, Chichester.

Badard, T. and Richard, D. (2001). Using XML for the exchange of updating information between geographical information systems. *Computers, Environment and Urban Systems*, **25**, 17–31.

Bao, S., Anselin, L., Martin, D. and Stralberg, D. (2000). Seamless integration of spatial statistics and GIS: The S–PLUS for ArcView and the S+Grassland Links. *Journal of Geographical Systems*, **2(3)**, 287–306.

Bernard, L., Schmidt, B., Streit, U. and Uhlenkuken, C. (1998). Managing, modelling, and visualizing high dimensional spatiotemporal data in an integrated system. *Geoinformatica*, **2(1)**, 59–77.

Bisby, F.A. (2000). The quiet revolution: biodiversity informatics and the Internet. *Science*, **289**, 2309–2312.

Bivand, R. and Gebhardt, A. (2000). Implementing functions for spatial statistical analysis using the R language. *Journal of Geographical Systems*, **2(3)**, 307–317.

Bivand, R.S. (2000). Using the R statistical data analysis language on GRASS 5.0 GIS database files. *Computers and Geosciences*, **26**, 1043–1052.

Blades, M. (1991). The development of the abilities required to understand spatial representations. In D.M. Mark and A.U. Frank, eds. *Cognitive and Linguistic Aspects of Geographic Space: An Introduction*, pp. 81–116. Kluwer Academic Publishers, Dordrecht.

Breunig, M. (1999). An approach to the integration of spatial data and systems for a 3D geoinformation system. *Computers and Geosciences*, **25**, 39–48.

Breunig, M., Cremers, A.B., Gotze, H.J., Schmidt, S., Seidemann, R., Shumilov, S. and Siehl, A. (1999). First steps towards an interoperable GIS: an example from Southern Lower Saxony. *Physics and Chemistry of the Earth, Part A: Solid Earth and Geodesy*, **24(3)**, 179–189.

Burnett, K., Bor–Ng, K. and Park, S. (1999). Comparison of the two traditions of metadata development. *Journal of the American Society for Information Science*, **50**, 1209–1217.

Cances, M., Font, F. and Gay, M. (2000). Principles of Geographic Information Systems used for Earth Observation data. *Surveys in Geophysics*, **21**, 187–199.

Choi, H., Kim, K. and Lee, J. (2000). Design and implementation of Open GIS component software. *International Geoscience and Remote Sensing Symposium (IGARSS)*, **5**, 2105–2107.

Clarke, K., Parks, B. and Crane, M. (2000). Integrating geographic information systems (GIS) and environmental models. *Journal of Environmental Management*, **59**, 229–233.

Clement, G., Larouche, C., Gouin, D., Morin, P. and Kucera, H. (1997). OGDI: toward interoperability among geospatial databases. *SIGMOD Record*, **26(3)**, 18–23.

Crosetto, M., Tarantola, S. and Saltelli, A. (2000). Sensitivity and uncertainty analysis in spatial modelling based on GIS. *Agriculture, Ecosystems and Environment*, **81**, 71–79.

Dollner, J. and Hinrichs, K. (2000). An object oriented approach for integrating 3D visualisation systems and GIS. *Computers and Geosciences*, **26**, 67–76.

Edwards, J.L., Lane, M.A. and Nielsen, E.S. (2000). Interoperability of biodiversity databases: biodiversity information on every desktop. *Science*, **289**, 2312–2314.

Egenhofer, M.J. (1999). Introduction: Theory and Concepts. In M. Goodchild, M. Egenhofer, R. Fegeas, and C. Kottman, eds. *Interoperating Geographic Information Systems*, pp. 1–4. Kluwer Academic Publishers, Boston.

Egenhofer, M.J., Glasgow, J., Gunther, O., Herring, J.R. and Peuquet, D.J. (1999). Progress in computational methods for representing geographical concepts. *International Journal of Geographical Information Science*, **13(8)**, 775–796.

Ehlers, M. (1997). Rectification and Registration. In J.L. Star, J.E. Estes and K.C. McGwire, eds. *Integration of Geographic Information Systems and Remote Sensing*, pp. 13–36. Cambridge University Press, Cambridge.

Faust, N.L. and Star, J.L. (1997). Visualisation and the Integration of Remote Sensing and Geographic Information. In J.L. Star, J.E. Estes and K.C. McGwire, eds. *Integration of Geographic Information Systems and Remote Sensing*, pp. 55–81. Cambridge University Press, Cambridge.

Fowler, C., Treml, E. and Smillie, H. (1999). Georeferencing the legal framework for a web based regional ocean management Geographic Information System. In *Proceedings of CoastGIS'99: GIS and New Advances in Integrated Coastal Management*, September 1999, Brest, France.

Gahegan, M. and Ehlers, M. (2000). A framework for the modelling of uncertainty between remote sensing and geographic information systems. *ISPRS Journal of Photogrammetry and Remote Sensing*, **55**, 176–188.

Gingras, F., Laks, V.S., Lakshmanan, I., Subramanian, N., Papoulis, D. and Shiri, N. (1997). Languages for multidatabase interoperability. *SIGMOD Record*, **26(2)**, 536–538.

Golledge, R.G. (1995). Spatial Primitives. In *Proceedings of the NATO Advanced Research Workshop: Cognitive Aspects of Human Computer Interaction for Geographic Information Systems*, March 1994, Palma de Mallorca, Spain.

Golledge, R.G. and Stimson, R.J. (1987). *Analytic Behavioural Geography*, p. 355. Croom Helms, Beckenham.

Goodchild, M.F. (2000). Foreword. In D. Wright and D. Bartlett, eds. *Marine and Coastal Geographical Information Systems*, pp. xiii–xv. Taylor and Francis, London.

Goodchild, M.F., Parks, B.O. and Steyaert, L.T. (1993). *Environmental Modelling with GIS*. p. 230. Oxford University Press, New York.

Guillaumont, B. and Durand, C. (1999). Integration et gestion de donnees reglementaires dans un SIG: analyse appliquee au cas des cotes Francaises. In *Proceedings of CoastGIS'99: GIS and New Advances in Integrated Coastal Management*, September 1999, Brest France.

Haining, R., Wise, S. and Ma, J. (2000). Designing and implementing software for spatial statistical analysis in a GIS environment. *Journal of Geographical Systems*, **2(3)**, 257–286.

Herring, J.R. (1999). The OpenGIS data model. *Photogrammetric Engineering and Remote Sensing*, **65(5)**, 585–588.

Hibbard, W. (1998). VisAD: Connecting people to computations and people to people. *Computer Graphics*, **32(3)**, 10–12.

Huang, B. and Lin, H. (1999). GeoVR: a web based tool for virtual reality presentation from 2D GIS data. *Computers and Geosciences*, **25**, 1167–1175.

Hwang, D., Karimi, H.A. and Byun, D.W. (1998). Uncertainty analysis of environmental models within GIS environments. *Computers and Geosciences*, **24**, 119–130.

Jensen, J.R., Cowen, D., Narumalani, S. and Halls, J. (1997). Principles of change detection using digital remote sensor data. In J.L. Star, J.E. Estes and K.C. McGwire, eds. *Integration of Geographic Information Systems and Remote Sensing*, pp. 37–54. Cambridge University Press, Cambridge.

Johnston, C.A., Cohen, Y. and Pastor, J. (1996). Modelling of spatially static and dynamic ecological processes. In M.F. Goodchild, L.T. Steyaert, B.O. Parks, C. Johnston, D. Maidment, M. Crane and S. Glendinning, eds. *GIS and*

*Environmental Modelling: Progress and Research Issues*, pp. 149–154. GIS World Books, Boulder.

Keller, S.F. (1999). Modelling and sharing geographic data with INTERLIS. *Computers and Geosciences*, **25**, 49–59.

Kingston, R., Carver, S., Evans, A. and Turton, I. (2000). Web based public participation geographical information systems: an aid to local environmental decision making. *Computers, Environment and Urban Systems*, **24**, 109–125.

Kottman, C.A. (1999). The Open GIS Consortium and progress toward interoperability in GIS. In M. Goodchild, M. Egenhofer, R. Fegeas and C. Kottman, eds. *Interoperating Geographic Information Systems*, pp. 38–54. Kluwer Academic Publishers, Boston.

Kraak, M.J., Muller, J.C. and Ormeling, F. (1995). GIS Cartography: visual decision support for spatiotemporal data handling. *International Journal of Geographical Information Science*, **9(6)**, 637–645.

Ladstatter, P. (2000). Interoperability and OpenGIS. *Geo Informations Systeme*, **13(1)**, 23–27.

Lees, J.M. (2000). Geotouch: software for three and four dimensional GIS in the Earth Sciences. *Computers and Geosciences*, **26**, 751–761.

Li, R. and Saxena, N.K. (1993). Development of an integrated marine information system. *Marine Geodesy*, **16**, 293–307.

Lloyd, R. (1997). *Spatial Cognition: Geographic Environments*, p. 221. Kluwer Academic Publishers, Dordrecht.

Lockwood, M. and Li, R. (1995). Marine Geographic Information Systems: what sets them apart. *Marine Geodesy*, **18**, 157–159.

Marble, D.F. (2000). Some thoughts on the integration of spatial analysis and Geographic Information Systems. *Journal of Geographical Systems*, **2(1)**, 31–35.

Mason, D.C., O'Conaill, M.A. and Bell, S.B.M. (1994). Handling four dimensional georeferenced data in environmental GIS. *International Journal of Geographical Information Science*, **8(2)**, 191–215.

McGwire, K. and Goodchild, M. (1997). Accuracy. In J.L. Star, J.E. Estes and K.C. McGwire, eds. *Integration of Geographic Information Systems and Remote Sensing*, pp. 110–133. Cambridge University Press, Cambridge.

Meaden, G.J. (2000). Applications of GIS to fisheries management. In D. Wright and D. Bartlett, eds. *Marine and Coastal Geographical Information Systems*, pp. 205–226. Taylor and Francis, London.

Morris, K., Hill, D. and Moore, A. (2000). Mapping the environment through three dimensional space and time. *Computers, Environment and Urban Systems*, **24**, 435–450.

Nyerges, T.L. (1994). Analytical Map Use. *Cartography and GIS*, **18**, 22–28.

Nyerges, T.L. and Golledge, R.G. (1997). NCGIA Core Curriculum in GIS, National Center for Geographic Information and Analysis, University of California, Santa Barbara, Unit 007, http://www.ncgia.ucsb.edu/giscc/units/u007/u007.html, posted November 12, 1997.

Paepcke, A., Cousins, S.B., Garcia–Molina, H., Hassan, S.W., Ketchpel, S.P., Roscheisen, M. and Winograd, T. (1996). Using distributed objects for digital library interoperability. *Computer*, **29(5)**, 61–68.

Phillips, A., Williamson, I. and Ezigbalike, C. (1999). Spatial data infrastructure concepts. *Australian Surveyor*, **44(1)**, 20–28.

Shyue, S. and Tsai, P. (1996). Study on the dimensional aspect of the marine Geographic Information Systems. *Oceans Conference Record (IEEE)*, **2**, 674–679.

Slater (1982). *Learning through Geography*, pp. 340. Heineman Educational Books, Ltd, London.

Su, Y. (2000). A user friendly marine GIS for multidimensional visualisation. In D. Wright and D. Bartlett, eds. *Marine and Coastal Geographical Information Systems*, pp. 227–236. Taylor and Francis, London.

Tschangho, J.K. (1999). Metadata for geospatial data sharing: a comparative analysis. *The Annals of Regional Science*, **33**, 171–181.

Valavanis, V., Georgakarakos, S., Koutsoubas, D., Arvanitidis, C. and Haralabus, J. (2002). Development of a marine information system for Cephalopod fisheries in the Greek Seas (Eastern Mediterranean). *Bulletin of Marine Science* (in press).

Wegner, P. (1996). Interoperability. *ACM Computing Surveys*, **28(1)**, 285–287.

World Wide Web Consortium (1998). Extensible Markup Language (XML) 1.0. W3C Recommendation, Feb. 1998. Available at: http://www.w3.org/TR/1998/REC-xml-19980210.

Wright, D. and Bartlett, D. (2000). *Marine and Coastal Geographical Information Systems*, p. 320. Taylor and Francis, London.

# CHAPTER TWO

# GIS and Oceanography

## 2.1 INTRODUCTION

Applications of GIS technology in the various oceanographic disciplines are multifaceted. They deal with data acquired with a variety of different methods and integration principles from different technological disciplines in an attempt to facilitate the resolution of the dynamics of the marine environment. Oceanographic GIS propose solutions to an ever-increasing number of marine problems and at the same time provide new insights to the unknown abyssal depths of our oceans. Ocean dynamics (oceanographic processes) as well as marine problems (e.g. conflicts in the coastal zone and pollution) have inherent spatiotemporal characteristics. Oceanographic data have two essential parts (location and attributes) that make them highly suitable for input to geospatial databases, which in turn open new ways for data storage, analysis and visualisation. The majority of oceanographic data often have a geographic location, are digital, uneven in distribution, originate from multiple institutions and can have resolutions that vary over many orders of magnitude. The spatial attribute of oceanographic data makes them highly suitable for GIS analysis as GIS provides a natural framework for the acquisition, storage and analysis of georeferenced data. GIS databases can store and manipulate several environmental parameters about ocean processes and present their effects on the marine environment in graphical format.

With GIS, the problems associated with simultaneously using multiple oceanographic datasets are handled efficiently and seamlessly. The computer-based GIS technology allows compact digital storage of information, which can then be maintained, updated, combined, generalised, reorganised and retrieved efficiently. The ease of spatial data manipulation using GIS allows the information to be viewed as a dynamic set of data with the static map being only a 'snapshot' of the database at a particular instance. Data layer combination allows decision makers to test multiple management scenarios and researchers to investigate data relationships using an iterative approach to refine research methods. In this chapter, current oceanographic GIS applications in coastal environments as well as in the open ocean are reviewed. GIS developments for the study of several oceanographic processes as well as for coastal mapping are examined. The importance of satellite imagery and the use of Internet technology to oceanographic and coastal GIS applications are also discussed.

Life in the ocean has an intimate relationship with the dynamics of ocean currents, which keep the ocean in a constant motion over an enormous range of scales, from microscale turbulence extending over a few millimetres to the great ocean gyres encompassing entire basins thousands of nautical miles in extent. Currents trigger certain physical oceanographic processes, which have a very significant biological importance. Biological productivity relies on nutrients being brought by currents to the surface from deep water and then the bloom of phytoplankton in the sea surface zone (euphotic zone), thus defining the seasonal state of the water masses. Mainly, such processes include upwelling, fronts, eddies, cyclonic and anticyclonic gyres, which are caused by the combination of earth's rotation, wind patterns and sea currents with coastal geomorphology and bottom topography playing a significant role to the spatiotemporal distribution and patterns of these processes.

The mixing of sea surface, intermediate and deep waters through seasonal nutrient concentrations, gyre formation and front systems, occurs on different spatial and temporal scales. These processes also vary in duration and strength. Their important effects in the state of water column and sea surface require monitoring and establishment of seasonality as well as measurement of their impact in productivity levels. Remote sensing (RS) of the marine environment comprises the greatest source of information about several parameters of the ocean surface while other sophisticated methods (e.g. moored and drifting buoy systems) provide information for the water column. The basic marine EO parameters are SST observed by infrared radiometers, ocean colour by spectrometers, sea surface elevation by altimeters and surface roughness by active and passive microwave systems that can be used to derive surface wind and waves. These data are invaluable for the study of ocean processes because they describe important parameters about these processes (temperature, CHL, currents, etc.) allowing a synoptic mapping of ocean processes in large sea areas. EO techniques routinely allow the production of quality products of ocean wind, waves, temperature, eddy and frontal location and propagation and water quality, such as CHL concentration and suspended sediments. For example, satellite imagery of SST and CHL concentration are very suitable for the study of such oceanographic processes. These processes can be actually identified on satellite imagery due to their difference in temperature and chlorophyll as compared to the surrounding area. Manipulation of such imagery within GIS can contribute to the indexing of these processes through different satellite data integration, providing information on the process duration, strength, epicentre location and direction, size and temperature/CHL variation. The increased attention to global environmental change and marine resource overexploitation are continually creating awareness of the potential of GIS technology for marine data analysis, data integration and creation of indexes on certain oceanographic processes (e.g. indexes on upwelling or major current systems).

Coastal populations worldwide are estimated to account for up to 60 per cent of Earth's total population establishing a significant portion of socio-economic activity in the coastal zone, where two thirds of the megacities are located. Almost 50 per cent of the US population live near the coast, as do 80 per cent of Australians and 45 per cent of the populations of countries bordering the Mediterranean Sea. The concentration of human population on the coast has risen through historical settlement, trading or political linkages, from the deterministic constraints of climate, availability of flat, fertile alluvial soils, proximity to fish stocks and for aesthetic and recreational reasons. Coastal waters serve as primary routes of transportation and communication among these population centres with more than 90 per cent of the pollutants generated by economic activities ending up in the coastal zone. Many marine problems that are now being encountered worldwide have resulted from the unsustainable use and unrestricted development of coastal areas and marine coastal and ocean resources. These problems include the accumulation of contaminants and pollutants in coastal and ocean areas, coastline erosion and the rapid decline of habitats and natural resources. In addition to the problems related to unsustainable marine exploitation, the marine environment is significantly affected by the impacts of human-induced climate change. The Intergovernmental Panel on Climate Change (IPCC) estimates that by the year 2100, the sea level will have risen by 31–110 cm. Wetlands are likely to be threatened, coastal erosion will increase and coastal resources, populations and economies will be adversely affected. In the shorter term, other effects of climate change such as changes in the frequency, intensity and patterns of extreme weather events like gales, intense precipitation and associated storm surges and flooding are likely to occur. Also, nutrient overload from sewage discharge, agricultural run-off, erosion and other land-based sources, has been identified by a UNEP Group of Experts on the Scientific Aspects of Marine Environment as the most widespread and serious cause of coastal pollution. Toxic chemical pollutants are another cause for concern. Harmful algal blooms disrupt marine life and toxic red tides threaten human life, while pathogenic bacteria and viruses create human health problems.

Despite the economic, social and ecological importance of the coastal zone, development and management of this inherently complex area is still largely pursued on a sector-by-sector basis and regulated on a jurisdictional basis. Examples of problems caused by this fragmented approach include habitat destruction, spatial conflicts and inefficient resource use. An integrated management approach becomes a necessity and requires coordination of those stakeholders whose actions significantly influence the quantity or quality of coastal resources and environments. Indeed, the need for an integrated management approach is widely acknowledged. Both UNCLOS and Agenda 21, from the UN Conference on Environment and Development held at Rio de Janeiro in 1992, call for integrated approaches to management. Nations have begun to respond to the call that 'each coastal state should consider establishing.... appropriate coordinating mechanisms....for integrated management and sustainable development of coastal and marine areas and their resources'. Canada's Oceans Act, for example, passed in January 1997, includes specific provisions for integrated coastal zone management and information-based management of ocean resources. In 1993, the Netherlands created the Coastal Zone Management Centre on behalf of the six Dutch ministries that shared jurisdiction of the coastal zone. The US National Oceanic and Atmospheric Administration (NOAA) created the Coastal Services Centre to provide 'an organisational focus for bringing together science, technology, and information to serve the needs of coastal resource managers, communities, and businesses'.

The implementation of effective information-based management of marine resources requires data collection and processing, including biological, social, economic and demographic data, analysis of natural systems, oceanographic processes and socio-economic activities, assessment of environmental, social and economic impacts of ongoing and/or proposed activities and evaluation of strategies. The GIS and RS are two technologies that have enabled integrated management by facilitating data collection, integration and analysis processes on these features. In parallel, the emergence of the Internet now makes resource databases accessible to a broader range of stakeholders. Data produced by a multitude of governmental agencies and non-governmental organisations have one value when used by the producing agency. However, when combined with other GIS data and connected to regulatory information, the value of these data is multiplied. Towards this goal, the combined use of GIS and the Internet greatly facilitate GIS analysed information dissemination among policy centres.

## 2.2 THE USE OF GIS IN VARIOUS FIELDS OF OCEANOGRAPHY

Oceanographic GIS are used in a variety of ways, such as data distribution tools, mapping tools and monitoring analysis tools and in a variety of disciplines, such as coastal zone assessment and management, ocean surface processes, marine geology and geomorphology, marine eutrophication, environmental and bioeconomic characterisation of coastal and marine systems, submerged marine habitats and marine habitat assessment, marine oil spill and pollution, ocean policy and management, climate change and sea-level rise, deep ocean mapping, flooding and natural hazard assessment and development of environmental sensitivity indices (ESI) maps. The latest GIS developments in oceanography include a variety of sophisticated approaches. For example, over the past decade, revolutionary technological changes have taken place in mapping and visualising the ocean floor. Multibeam sonar systems, which use beam-forming techniques to create large swaths of the seafloor, produce high-resolution (both lateral and vertical) bathymetry and seafloor imagery (from acoustic backscatter). This technology produces large amounts of data (in density and volume) presenting fundamental challenges in terms of data interaction, integration and interpretation. A developing solution to these challenges is the use of GIS for the storage, interpretation and visualisation of multibeam sonar data. Any point in such datasets can be interrogated for position, depth, or any other attribute. Through artificial sun illumination, shading and 3D rendering, massive digital bathymetric data sets can easily be explored in the form of a natural looking and easily interpretable landscape. Spatial measurements can be made and certain profiles extracted while datasets can be maintained at their original resolution.

Following, a review of scientific literature on the use of GIS in various oceanographic disciplines and themes is presented revealing the great diversity of GIS applications in oceanographic studies.

### 2.2.1 Marine Geology

The use of GIS technology in marine geology brought new insights into the organisation and analysis of marine geological data, which resulted in the mapping of abyssal environments and processes, which until recently were unknown. Kunte (1995) reviewed 7500 worldwide geological databases, found only 110 databases related to marine geology, 71 marine geology databases numeric in nature and proposed that GIS should be utilised for the georeferencing of these databases. New information on several deep ocean areas is made available. Wright (1996) used GIS to interpret data collected by the famous Alvin submersible when it lowered from RV Atlantis II in an expedition to the Juan de Fuca Ridge in the Pacific Ocean. Smith and Sandwell (1997) created a high-resolution digital bathymetric map of the oceans by combining available depth soundings with high-resolution marine gravity information from the GEOSAT and ERS-1 spacecrafts. The map shows new deep ocean features (e.g. seamounts chain in the South Pacific) and relations among the distributions of depth, sea floor area and sea floor age. Bobbitt *et al.* (1997) integrated multidisciplinary oceanographic data from the eastern North Pacific Ocean (sonar hydrophone array data, multibeam sonar bathymetry, satellite RS and field observation data) and applied geophysical GIS analysis for seismicity research. Goldfinger and McNeill (1997) and Goldfinger (2000) discussed methods of active tectonics data acquisition, analysis and visualisation using GIS. Tsurumi (1998) illustrated the use of GIS as a supplementary tool for analysis of temporal change and succession in biological systems at the hydrothermal vents in North Cleft Segment, Juan de Fuca Ridge. Wright (1999) reviewed the newest advances in mapping and management technologies for undersea geographic research, particularly on the ocean floor and illustrated the important contribution of geographers to the science of the ocean floor through the use of GIS for handling data from the deep ocean.

Livingstone *et al.* (1999) mentioned the suitability of GIS as an integration tool in coastal geomorphology studies using airborne data, derived from video and digital camera imagery, with terrain data from ground surveys. The combination of such imagery with terrain data in GIS consists a marine geologist's tool either for data visualisation and interpretation or for feature definition and measurement of change. Lee *et al.* (1999) applied GIS spatial analysis techniques for evaluation of slope stability in an effort to determine the locations of submarine slope failures. Regional density, slope and level of anticipated seismic shaking information were combined in a GIS framework to yield maps that illustrate the relative stability of slopes in the face of seismically induced failure. Swath bathymetric mapping showing seafloor morphology and distribution of slope steepness and sediment analysis of box cores were integrated to delineate the variability of sediment density near the seafloor surface. Wong *et al.* (1999) combined data from sidescan sonar, multibeam bathymetry, physical samples, current moorings and photographs of the sea bottom in a GIS to reveal new details about the geology, morphology and active geologic processes of the Monterey Bay National Marine Sanctuary. Pratson *et al.* (1999) merged the US Geological Survey's Digital Elevation Models with a vast compilation of hydrographic soundings collected by the National Ocean Service and various academic institutions to produce a land and seafloor elevation database for US coastal zone including the 200-mile offshore limit. The database and a set of tools were placed on a CDROM allowing users to view grid images and modify the elevation grids for import into GIS applications. Ganas and Papoulia (2000) used a combination of image processing and GIS techniques connected to a model that calculates probabilistic estimates of ground motion parameters to develop high-resolution seismic hazard maps for an underwater rift in central Greece. McAdoo *et al.* (2000) applied GIS to study submarine processes, including sediment transport mechanisms and slope stability. They measured various aspects of submarine landslides including landslide area, runout distance and headscarp height along with the slope gradient of the runout zone, the failure's scar, headscarp and adjacent slopes. Wright *et al.* (2000) produced a series of GIS maps for the Tonga Trench in Western Pacific by integrating bathymetric data from a multibeam swath mapping system, seismic reflection data, gravity data and total field magnetic data.

### 2.2.2 Flood Assessment

Flood research using GIS is also extensive following various approaches in which satellite imagery, risk assessment models and digital elevation models (DEM) are often integrated. Scholten *et al.* (1998) proposed a GIS infrastructure among flood control emergency managers in the Netherlands and the development of integrated software tools with input from several models for the detection of weak parts in dikes and generation of evacuation plans. Blomgren (1999) used kriging in GIS to produce a high-resolution DEM of the low-lying Falsterbo Peninsula in south Sweden. Clustered input points were removed, significantly improving the agreement between the actual and theoretical variograms. The DEM was used to visualise flooding scenarios predicted to occur more frequently in the future due to the increased greenhouse effect. Yang and Tsai (2000) presented the GIS-based Flood Information System (GFIS) for flood plain modelling, flood damages calculation and flood information support. The primary advantages of GFIS are the abilities to accurately predict the locations of flood area, depth and duration, calculate flood damages in a flood plain and compare the reduction of flood damages for flood mitigation plans. Nico *et al.* (2000) used amplitude change detection techniques for flood area detection in southern France from multipass Synthetic Aperture Radar (SAR) data. Coherence derived from multipass SAR interferometry was used as an indicator of changes in the electromagnetic scattering behaviour of the surface, thus potentially revealing all the areas affected by the flood event at any time between the two passes. The output map is in a format that can be further used by most commercial GIS packages. Islam and Sado (2000) developed an algorithm for classifying cloud-covered pixels in AVHRR images into water or non-water categories. The purpose of this algorithm is to help in flood hazard assessment in Bangladesh where cloudy skies often occur during big floods. Classified images are combined with land cover and elevation data in a GIS for the production of flood hazard maps.

Thumerer *et al.* (2000) developed a GIS tool to determine coastal vulnerability to flooding along the English east coast. Results from oceanographic and climatic research were combined with data on sea defences, elevation values and patterns of landuse. A risk assessment model was integrated to estimate flood return periods according to different climate change scenarios for the years 2050 and 2100. Dobosiewicz (2001) integrated a DEM and several flood zone datasets for a New Jersey urban estuary (Raritan Bay) to identify land features within 1 per cent probability flood zones. Berz *et al.* (2001) used GIS to integrate results of earlier work with latest findings in scientific literature for the creation of the third edition of the World Map of Natural Hazards. In the production of this third edition, all the basic data were for the first time recorded, adjusted and analysed in GIS. Results include a hardcopy map, a 30 cm globe and a CDROM version, which were produced exclusively with the techniques of digital cartography. Hazard information in these products include earthquake and volcanism, windstorms, floods, marine hazards and effects of El Nino and climate change with three essential components: (1) intensity; (2) frequency; and (3) reference period.

### 2.2.3 Coastal and Ocean Management

The use of GIS in coastal and ocean management is continually growing through a diverse number of applications and initiatives. This field of GIS applications is very important because it deals with areas that are characterised by a diverse number of spatial conflicts, currently managed under a fragmented jurisdictional frame. The compilation of a diverse number of datasets under GIS frameworks provides a common source of information to the various coastal management authorities. Beusen *et al.* (1995) developed a GIS-based model that integrates geohydrological data for the estimation of loads of nitrogen and phosphorus in the coastal seas of Europe. In a similar study, Nasr *et al.* (1997) related sediment distribution to heavy metal concentrations and pesticides in Abu Qir Bay in Egypt. Smith and Lalwani (1996) discussed the important role GIS plays to the cooperation of organisations in various international agreements for the sea use management in the North Sea including maritime transport, fisheries, waste disposal, recreation, research and conservation. Chua (1997) noted that science plays a significant role in the management of coastal and marine areas particularly in providing the scientific basis for policy interventions through the development of integrated products that facilitate the defective communication between scientists and decision makers, intellectual and cultural arrogance and inadequate technical and management capability at the local level. Fowler and Schmidt (1998) identified that the coastal and ocean resource management community has been slower than that of land-based management to develop GIS capacity as a management tool. Various GIS products from NOAA's Coastal Services Centre are developed to bridge the gap between coastal science and coastal management in a framework of data compilation and delivery, application development, training and policy/GIS integration (http://www.csc.noaa.gov/).

Basu (1998) created a coastal GIS for the Island of Oahu (Hawaii) by integrating various land and marine data including bathymetry, gravity and magnetic data from the National Geophysical Data Centre (NGDC) database, GLORIA sidescan sonar data collected by the US Geological Survey (USGS), coastline data from Global Self consistent Hierarchical High resolution Shoreline (GSHHS) database, SPOT satellite data and elevation data. Spatial analysis among these different datasets produced 3D surfaces of the area. Li *et al.* (1998) created a monitoring and management GIS for Malaysia's shoreline. The tool includes spatial data (shoreline locations, topographic data, bathymetric data, parcel data, buoy locations), time series data (wind and wave observations), social and economic data and aerial photographs. The coastal GIS consists of three subsystems (shoreline erosion monitoring, coastal engineering management and coastal data inventory). The project was supported by the Asian Development Bank and conducted by an international and interdisciplinary team. Van Zuidam *et al.* (1998) presented a research programme of the International Institute for Aerospace Survey and Earth Sciences (ITC), in which RS, GIS, modelling and *in situ* measurements were used for the development and evaluation of scenarios for coastal zone management. Methodologies included hypothesis generation based on optimum RS datasets, parameter estimation, evaluation and validation and prediction of the physical aspects of coastal landscape development under the influence of natural processes and human impacts.

Aswathanarayana (1999) proposed a Natural Resources Management Facility (NRMF) for Mozambique to achieve poverty reduction and employment generation through ecologically sustainable, economically viable and people participatory management of natural resources (water and soils, coastal and marine, ecotourism, energy, mineral). The proposal included a comprehensive GIS database for all available data to obtain a synoptic picture of the environmental situation in the country. The NRMF is concerned with coastal resources management, which includes the improvement of the productivity of coastal ecosystems, the prevention of marine pollution, the protection of ecosystems like mangroves and corals and preservation of the quality of seafood. West (1999) developed a GIS-based decision support system for South Florida's coastal resources. The system uses a variety of datasets, GIS analysis and other embedded modelling and features a user interface for dynamic interaction. The tool was build for the Florida Department of Community Affairs to support a diverse set of decisions made by policy officials involved with Florida's coastal resources. Yang *et al.* (1999) identified that due to the large magnitude and rapid rates of change in deltaic lowlands, spatial surveillance systems are needed to efficiently measure and monitor channel migration. They developed a GIS for the study of channel migration in the highly active Yellow River Delta in China using a series of time sequential Landsat images spanning a period of 19 years to systematically examine the spatiotemporal changes of river banks and channel centrelines and relate these computational results with appropriate natural and human processes affecting the delta. Their method might be used in mapping highly dynamic deltaic environments, which is a complicated task due to the rapid and significant changes of many fluvial forms and lowland coasts and the poor topographic expressions of the landforms, which in turn makes updating of landform maps crucial to environmental management and development. Neilson and Costello (1999) introduced basic information on seabed types available on marine charts of Ireland to label in GIS the high-tide mark shoreline according to the corresponding dominant seabed types. This approach is applicable to any coastal area for which basic seabed information exists and it can be used as basic information for management and research purposes. Stanbury and Starr (1999) developed a GIS for the Monterey Bay National Marine Sanctuary that allows manipulation of many terrestrial and marine datasets (land cover classification, benthic habitat types, fisheries, watersheds) aiming to create a broad spatiotemporal database for the evaluation of natural resources, permitting and monitoring coastal developments and assessing environmental impacts. Capobianco (1999) and Belfiore (2000) discussed the important role of new technologies in coastal management in Europe noting that many European Communities (EC) funded demonstration projects have reported the successful use of GIS for resource inventories, analysis and monitoring. Garcia *et al.* (2000) incorporated four physical (lithology, landforms, river discharge and marine processes) and two anthropomorphic (population growth and urbanisation) components in a GIS to develop a sensitivity index for the coastline of the Costa del Sol, southern Spain. MacDonald and Cain (2000) used GIS to rank the environmentally sensitive areas of the UK coastline based on assessments of pollution risks associated with shipping (description of different hazards, vessels routing patterns and historical frequencies of shipping accidents). Leshkevich and Liu (2000) presented the Great Lakes CoastWatch regional Internet node (http://coastwatch.glerl.noaa.gov), a real time data delivery interactive GIS system that allows viewing and analysis of satellite surface temperature and visible imagery with data overlays such as bathymetry, gridded wind fields and marine observation data. Edgar *et al.* (2000) underlined that estuaries, the most anthropogenically degraded habitat type on earth, must be categorised into particular types and levels of human impact. They used GIS to categorise Tasmania's 111 large and moderate-sized estuaries based on environmental factors and the state of estuarine biodiversity. Yetter (2000) used frequency distribution analysis in GIS to develop environmental indicators for Delaware's coastal zone aiming to allow for more scrutiny in examining management efforts. Li *et al.* (2001) applied various spatial modelling and analysis methods in high-resolution imagery to detect shoreline changes along the south shore of Lake Erie. The shoreline is represented as a dynamically segmented linear model that is linked to a large amount of data describing shoreline changes. Klemas (2001) reviewed the latest advances in RS data use through GIS for estimating coastal and estuarine habitat conditions and trends (watershed land cover, riparian buffers, shoreline and wetland changes). Latest advances in the application of GIS help to incorporate ancillary data layers to improve the accuracy of satellite imagery classification providing coastal planners and managers a means for assessing the impacts of alternative management practices.

### 2.2.4 Coastal Zone Dynamics

The modelling of the dynamics of the coastal zone is another common GIS application especially when combined with associated hydrodynamics and morphodynamics models. Tortell and Awosika (1996), on an UNESCO manual on oceanographic survey techniques and living resources assessment methods, mentioned the importance of GIS as an excellent tool for coastal area planning and management and an efficient mechanism for handling data and transforming it into information. Bettinetti *et al.* (1996) used GIS to develop a tool for the design of practical measures to be implemented within the Venice Lagoon boundaries in Italy. The developments included a description of the lagoon ecosystem, identification of the main aspects of the degradation, identification of critical environmental parameters and definition of target conditions for recovery plans, verification of the selected target conditions and identification of critical areas and evaluation of intervention alternatives in relation to their technical and economical feasibility and their socio-economic impact. This analytical approach resulted in long-term measures for the reduction of pollution inputs from the drainage basin, medium-term measures for the modification of the morphology and of the water circulation pattern in specific areas and emergency actions for macroalgae harvesting. Hesselmans *et al.* (1997) discussed the possibilities of RS technologies as these apply in the hydrodynamics and morphodynamics of the coastal zone noting that an integrated approach combining RS data and specific hydrodynamic modelling is highly suitable for obtaining quantitative information on determining coastal process parameters. Lin *et al.* (1999) integrated a GIS and an ocean fluid dynamics model to describe the dynamic tide system of the East China Sea. They showed how the tide system is affected by sea bottom friction, underwater topography and shoreline morphology influencing the formation of large-scale sand ridge clusters. Crowley *et al.* (1999) used a real time RS and *in situ* observation network to study certain oceanographic processes, such as upwelling. The data network initialised GIS overlays among satellite derived SST and CHL concentration, surface currents, meteorology and AUV subsurface temperature, salinity and current profiles. Spaulding *et al.* (1999) developed a GIS-based Water Quality Mapping and Analysis Programme (WQMAP) for modelling the circulation and water quality of estuarine and coastal waters. WQMAP addresses the problem that GIS generally have on the production of high-quality animations when large amounts of time varying data are used by providing a stand alone application with simplified interactive GIS functionality and a highly optimised viewer for animation of temporally and spatially varying model results. It also has the option to export model results efficiently to external GIS applications for further analysis. WQMAP includes three basic components (a boundary fitted coordinate grid generation module, a 3D hydrodynamics model and three separate water quality or pollutant transport and fate models).

### 2.2.5 Marine Oil Spills

GIS applications, developed for the study of marine oil spill pollution, include several sophisticated approaches, as well. GIS provide the tools for the development of oil spill sensitivity indices and when combined with satellite images and spill propagation models contribute to the identification and evolution of oil spill pollution. Davis *et al.* (1994) discussed the digital merging of same day SPOT panchromatic and mutlispectral imagery for the creation of a sophisticated oil spill GIS modelling programme. Smith and Loza (1994) outlined the efforts of the Texas General Land Office (GLO) for response to oil spills and particularly the GIS developments carried by GLO to assist in oil spill related research. Populus *et al.* (1995) evaluated the suitability of various satellite sensor data (SPOT, Landsat Thematic Mapper and ERS-1 Radar) as data input to GIS for the study of environmental sensitivity to oil pollution. Krishnan (1995) demonstrated the efficiency and effectiveness of GIS in the identification of critical areas that need to be protected in the event of an oil spill in the Shetland Islands, United Kingdom. Douligeris *et al.* (1995) presented an integrated information management tool consisting of an object relational database management system, an intelligent decision support system, an advanced visualisation system and a GIS for handling large and diverse databases of environmental, ecological, geographical, engineering and regulatory information for risk analysis and contingency planning. Sorensen (1995) discussed the design and development of a commercial GIS-based oil spill contingency planning and response prototype application. Rymell *et al.* (1997)

implemented three modelling systems under a common GIS database and user interface. The system serves as an assessment tool of the contamination of seabed by drilling operation discharges and the potential for accidental oil spillage in the Isle of Man. Thia-Eng (1999) provided a brief analysis of a project on marine pollution from land and sea sources in East Asian Seas, in which experts from Malaysia, Indonesia and Singapore focused on the compilation of Malacca Strait Environmental Profile, a GIS-based interactive management atlas. Moe *et al.* (2000) developed a GIS implemented model, which is initiated by geomorphologic maps and georeferenced biological data and characterises the degree of habitats' oil sensitivity and vulnerability. The approach is based on the fact that historical oil spills have shown that environmental damage on the seashore can be measured by acute mortality of single species and destabilisation of the communities. Li *et al.* (2000) underlined that spatial data quality has an impact on coastal oil spill modelling and pointed to a reconsideration of the coupling strategies for GIS and environmental modelling, which should be exposed to include specific spatial data quality analysis tools. Muskat (2000) presented GIS developments by the Office of Spill Prevention and Response (Department of Fish and Game, California) for oil spill preparedness, during an emergency response and as an aid for quantifying natural resource damage. Ducrotoy *et al.* (2000) discussed marine pollution issues in the North Sea and the use of GIS in pollution modelling in the area.

Crane *et al.* (2000) developed a GIS to address the radionuclide contamination issue in the Arctic Ocean. Information released in 1992 on deliberate dumping of nuclear materials (including 16 nuclear reactors, six of them with fuel rods intact and over 10 000 containers of lower level radioactive waste) in shallow Siberian and Arctic Seas elicited a strong response among the countries ringing the Arctic. The GIS tool includes heavy metal and organochlorine contamination data, compiled for the period from 1960 to 1997, resulting in the Arctic Environmental Atlas. A series of contamination maps and dumping locations are available through this atlas. Stejskal (2000) described a GIS system that is used by the Australian Government for evaluation of the risk of drilling projects on adjacent to sensitive resources, such as coral reefs and mangroves and for government approval purposes. The system integrates a suite of algorithms that predict the movement and weathering of oil, using the chemical composition of specific oil types. Zheng *et al.* (2001) analysed Earth photographs, acquired by Space Shuttle Columbia, for the observation of oil slicks in the northern Arabian Sea. The length and width of oil slicks as well as the total volume of oil discharged into the sea were estimated. Tsanis and Boyle (2001) developed a 2D hydrodynamic pollutant transport model, which is linked to a GIS for data input, processing and output of spatially distributed information.

### 2.2.6 Sea-Level Rise

Modelling of the various environmental and social consequences from sea-level rise due to global warming is also a common GIS task. Modelling and mapping the consequences of sea-level rise to coastal areas are important factors to short-term coastal protection and long-term coastal planning. Zeidler (1997) developed a GIS-supported coastal information and analysis system to support coastal zone management in relation to climate change and accelerated sea-level rise. Four sea-level rise scenarios (10 cm and 30 cm by 2030, and 30 cm and 100 cm by 2100) have been assumed as boundary conditions for the entire coast of Poland and three adaptation strategies (coast retreat, limited protection and full protection) have been adopted and compared in physical and socio-economic terms. El-Raey *et al.* (1997) and El-Raey (1997) made a quantitative assessment of the vulnerability of the Nile delta coast of Egypt to the impacts of sea-level rise using GIS and RS techniques together with ground-based surveys. The vulnerability assessment study showed that over two million people will have to abandon their homes and the loss of the world famous historic, cultural and archeological sites will be unaccountable due to a sea-level rise of 50 cm. King (2000) presented methodological aspects of using GIS in vulnerability assessment, economic valuation and coastal planning in relation to the effects of global climate change (sea-level rise) in 12 Caribbean countries. Hennecke *et al.* (2000) used GIS for the modelling of near future sea-level rise potential impacts in south-east Australia. Greve *et al.* (2000) integrated property information data and hazard information from local government agencies to develop a formal GIS method and map a generalised risk zone on two New South Wales open ocean beaches representing possible present and future areas at risk, weighted in terms of hazard level and exposure values and to quickly assess areas at risk. Jones (2000) used the GIS-based sea-level affecting marshes model (SLAMM4) to detect migrating shorebird habitat loss from sea-level rise due to global climate change at several key coastal sites, which birds use as their staging sites. Young and Bush (2000) underlined that GIS is an important tool in hazards assessment consisting the state of the art in coastal hazards mapping, research and policy.

Simas *et al.* (2001) combined ecological modelling with GIS analysis and RS to determine the effects of sea-level rise in estuarine salt marshes. Several sea-level rise scenarios are generated to determine changes in global salt marsh productivity showing that mesotidal estuaries are susceptible to sea-level rise only in a worst case scenario, which is more likely to occur if the terms set out by the Kyoto protocol are not met by several industrialised nations. Titus and Richman (2001) published various maps showing lands vulnerable to sea-level rise along the US Atlantic and Gulf Coasts. The production of these maps offers a first order examination of the vulnerability to the rise in sea-level expected to result from global warming, predicting of lands vulnerable to coastal storm surge and state- or local-level planning considerations of long-term sea-level rise in the coastal zone.

### 2.2.7 Natural and Artificial Reefs

Natural reefs, as important ecosystem health indicators, and artificial reefs, as mechanisms of biodiversity sustainability, are also studied with the use of GIS. Mapping and protecting natural reefs and developing artificial reefs became worldwide important issues for protecting biodiversity and sustaining fish populations. Spalding and Grenfell (1997) presented a global assessment of the total area of coral reefs using maps of reef areas digitised in GIS while Kleypas *et al.* (2000) integrated global environmental data of SST, salinity, water depth, water clarity and solar irradiance for predicting worldwide reef occurrence. Wright *et al.*

(1998) used GIS to identify optimal locations for the possible sitting of artificial reefs in areas threatened by pressure of industry, shipping and recreational activities. The GIS was used to a feasibility study of constructing a steel artificial reef in Moray Firth using decommissioned platforms from the oil and gas industry in the North Sea. The criteria for site selection were divided into broad constraints, such as 'outside the main trawling areas and shipping lanes' and more narrowly defined criteria, such as 'away from locations of recreational centres along the coast, but not too far offshore', 'at least 40 m water depth and level seabed topography' and 'locations of medium or fine compacted sand'.

Shafer and Benzaken (1998) identified that wilderness, a traditionally terrestrial resource designation, should be also applied to marine settings (marine wilderness) and proposed that characteristics of wilderness related to human presence, natural features and remoteness must be taken into account in management methods derived by the GIS-based Australian National Wilderness Inventory as these apply to the Great Barrier Reef National Park (NE Australia). Johnson *et al.* (1999) utilised RS and GIS to study the effects of anthropogenic change in 31 adjacent river catchments that drain directly into the waters of the Great Barrier Reef. Caldwell (1999) presented the Marine Ecosystem Geographic Information System (MEGIS) working group, which is a collaboration of various agencies across the Pacific for the understanding, managing and preserving of coral reef ecosystems of the islands in the Caribbean and Pacific. The World Conservation Monitoring Centre (WCMC) prepared ReefBase, a user-friendly database on coral reefs and their resources. Information gathered from the published literature as well as conference proceedings, technical reports, news-articles and various theses and manuscripts resulted in the development of two mapping systems, WinMap and ReefMap, which are incorporated into ReefBase. ReefMap displays standard format maps of the major world reef systems while WinMap provides geographic displays of data included in ReefBase (http://www.reefbase.org/).

## 2.2.8 Wetlands and Watersheds

Wetland and watershed research using GIS includes a variety of developments targeting to the restoration management and sustainable use of these resources. Jansson *et al.* (1998) focused on the eutrophication problem of the Baltic Sea, the use of wetlands as natural nutrient sinks and developed a grid cell-based GIS approach to analyse the nitrogen retention capacity of natural wetlands in the Baltic drainage basin. Semlitsch and Bodie (1998) used GIS to question wetland regulations, drafted by the US Army Corps of Engineers, which included that small and isolated wetlands are likely, continue to be lost. They showed that small wetlands are not expendable, if biologically relevant data on the value of such wetlands is available. Olsvig-Whittaker *et al.* (2000) used GIS to generate, evaluate and compare different conservation management scenarios, derived from field data, for a wetland nature reserve in Israel. Cedfeldt *et al.* (2000) developed a GIS-based wetland assessment method that enumerates spatial predictors for three primary wetland functions (flood flow alteration, surface water quality improvement and wildlife habitat). Using remotely sensed land use information and DEMs, this method produces three separate grids of wetlands that perform each function. The method was tested on four watersheds in Vermont's Lake Champlain Basin. Results and preliminary verification indicate that the method can successfully identify those wetlands in the US northeast region that have the potential to be functionally important. Aspinall and Pearson (2000) linked landscape ecology and environmental modelling with GIS to provide an integrated tool for geographical assessment of environmental condition in water catchments. They combined a series of indicators of water catchment health representing state and trend, focusing on their physical, biological and chemical properties as well as their overall ecological function. Wang (2001) demonstrated the usefulness of GIS for the study of the spatial relationships between land uses and river water quality measured with biological, water chemistry and habitat indicators in watersheds, thus integrating water quality management and land use planning.

### 2.2.9 Submerged Aquatic Vegetation

Submerged vegetation, as important substrate especially during the early life of many aquatic species, is also examined through GIS techniques. A few studies are presented here touching this field of GIS applications later in Section 2.11. Rea *et al.* (1998) measured the relative effects of several physical factors (water depth, sediment slope steepness, aspect and fetch) on the development of macrophytes in a pond. They scanned a series of aerial photographs for the period 1979–1992 and used GIS to classify the photographs and compare the rates and patterns of macrophyte development across years. DeAngelis *et al.* (1998) developed an ecosystem landscape model, which uses GIS vegetation data and existing hydrology models for South Florida Everglades Ecosystem to simulate and compare the landscape dynamic spatial pattern of several wildlife species resulting from different proposed water management strategies. De Jonge *et al.* (2000) presented a restoration strategy, which contains a selection procedure for suitable transplantation sites of eelgrass in the Dutch Wadden Sea, integrating in a GIS model factors such as sediment composition, exposure time, current velocity, wave energy and tidal depth.

### 2.3 WORLDWIDE OCEANOGRAPHIC GIS INITIATIVES

Several oceanographic initiatives include GIS as data management and analysis tool, a practice often resulting in the development of marine oceanographic atlases for specific areas that are distributed in the form of CDROM or Internet GIS databases. For example, the Persian Gulf States in general and Kuwait in particular are facing many socio-economic challenges that are caused by several environmentally related issues. For this purpose, scientists at the Kuwait Institute of Scientific Research created the GIS-based Environmental Information System (EIS), a comprehensive system with a vast amount of terrestrial, atmospheric and marine data capable of integrating environmental data with relevant socio-economic issues to support the decision and policy makers in achieving the nation's goals of sustainable development. Kuwait's EIS includes classification of geographical datasets for coastal zone sensitivity index maps and for several oceanographic characteristics (Al-Ghadban 1997).

In the United States, GIS was chosen as a solution to the increasing data volume and complexity of the various Vents Programme datasets. Started in early 1980s, NOAA's Pacific Marine Environmental Laboratory (PMEL) Vents Programme resulted in various geolocated datasets from a wide range of disciplines, including geophysics, geology, physical and chemical oceanography and acoustic monitoring and deepsea biology. The Vents Programme GIS brought multidisciplinary data under a single hardware and software platform and provided overlay, analysis and query capabilities for the complex multidisciplinary datasets (Fox and Bobbit 2000). In Europe, ERGIS (European Marine Resource Geographical Information Service), coordinated by the Marine Information Service (MARIS) in the Netherlands, embraces the vast range of subjects (form oceanography to shipping lanes), which describe the ecostructure and exploitation of coastal and offshore waters. Features of the natural environment such as coastlines, bathymetry, seabed morphology, mineral reserves, geology and geotechnics, can be combined with constructs such as environmental sites and fishing grounds as well as man made features and seabed obstacles. Through a highly user-friendly client system, ERGIS provides access to a comprehensive database on marine installations, bathymetry, seabed morphology, oceanography and commercial and marine activities. GIS functionality permits the manipulation of data on the basis of spatial relationships (http://www.maris.nl/).

Scientists at the Mapping and Geographic Information Centre (MAGIC) of the British Antarctic Survey (BAS) developed image analysis and photogrammetric techniques and utilised remotely sensed data for the preparation of the Antarctic Digital Database (ADD), a collaborative international project. The first digital topographic database of Antarctica, published on CDROM in 1993 under the auspices of the Scientific Committee on Antarctic Research (SCAR), was a milestone in Antarctic mapping. Now, a third web-based version of ADD provides the common framework to which multidisciplinary datasets can be referred for GIS applications, facilitating retrieval and evaluation of data, particularly for monitoring environmental changes in Antarctica (http://www. nerc-bas.ac.uk/public/magic/add_home.html). In addition, GIS was used for the generation of an Antarctic DEM integrating cartographic and remotely sensed data. The Antarctic DEM provides exceptional topographical details and represents a substantial improvement in horizontal resolution and vertical accuracy over the earlier, continental scale renditions, particularly in mountainous and coastal regions (Liu *et al.* 1999). Multiyear participation of the Georgia Geologic Survey (GGS) in the Minerals Management Service funded Continental Margins Programme included geological and geophysical data acquisition through several offshore stratigraphic framework studies (phosphate bearing Miocene age strata, distribution of heavy minerals, near surface alternative sources of groundwater) and the development of a coastal GIS focusing on investigations of economic minerals on the Georgia coast (Cocker and Shapiro 1999). The US Naval Oceanographic Office has developed a Survey Planner and Status Tracking System prototype, which includes custom GIS extensions that allow the tracking of ship locations and monitoring of their status in near real-time. The system also manages vast amounts of information from survey data (bathymetric, hydrographic, oceanographic and geophysical) collected by vessels at sea (Mesick *et al.* 2000). An initiative is currently being taken by several Norwegian organisations entitled 'MAREANO: Marine Areal Database for the Norwegian Sea' for the mapping of marine sea floor off Norway. The project investigates a commercially important region for fisheries and the petroleum industry and includes the world's largest system of cold-water coral reefs. The obtained information will be stored in a GIS database and will be available to environmental managers and interest groups as well as the fisheries, aquaculture and petroleum industries (Noji *et al.* 2000).

## 2.4 ONLINE OCEANOGRAPHIC GIS TOOLS

Oceanographic GIS development on the Internet provides a large number of benefits, such as increased information provision and accessibility and enhanced communications and networking within the community. Thus, in connection with the Internet technology, GIS developments are widely expanded falling into four main categories. The first concerns metadatabases, where the systems available provide the pointers to databases. The second category concerns multidimensional databases, where online data servers distribute extended marine datasets. The third category includes applications specifically developed for certain marine areas and are used by several management authorities as a source to common oceanographic information. The fourth category includes communication systems, which serve as a platform for the community to exchange ideas and information. For example, the Ocean Planning Information System (OPIS) developed by the NOAA Coastal Services Centre (CSC), in cooperation with the states of North and South Carolina, Georgia and Florida is the first attempt in the United States to create a regional, multistate information system for the coastal ocean. The system is based on a GIS database with regional georeferenced regulatory and environmental spatial data and provides management scenarios through an online mapping tool (http://www.csc.noaa.gov/opis/). Generally, the need to interchange information among scientific analysis and policy development in local and global scales became a necessity since marine problems are beyond any boundary and of international common concern.

The Internet, as a data communication platform among scientific communities and policy makers, is widely utilised by several organisations, which have already developed and maintain GIS monitoring tools for many sensitive marine areas worldwide and disseminate common information among various management authorities. Several web-based GIS disseminate both GIS ready data and GIS analysed results for specific areas. A short description of these applications is provided below:

- Aral Sea (http://giserv.karelia.ru/aral/): The Aral Sea GIS was developed through cooperation among the German Aerospace Centre, Nukus State University, the Russian Academy of Sciences and the Uzbek government and was highly supported by the World Bank, UNDP, USAID and the European Communities. Aral Sea GIS covers the Aral Sea, AmuDarya Delta and SyrDarya Delta in Uzbekistan and Kazahstan, both Republics of Central Asia.

The main purpose of the system is to develop models of optimal water and land use of irrigated lands minimising ecosystem transformation and desertification.

- Baltic Sea (http://www.grida.no/baltic/): The Baltic Sea Region GIS is a widely used resource of administrative units, arable and pasture lands, coastline, land cover, population density, subwatershed drainage basins and wetland distribution for the Baltic region (Jansson *et al.* 1999). Besides professionals in the Baltic environmental community, many educational institutions and policy offices are taking advantage of this resource.
- Barents Sea (http://www.nodc.noaa.gov/OC5/barsea/barindex1.html): The Climatic Atlas of the Barents Sea is a CDROM product and an online mapping tool developed by the Russian Academy of Sciences and the World Data Centre-A (Oceanography). More than 74 000 sampling stations within the Barents Sea covering the period 1898–1993 were inserted in the CDROM.
- Bering Sea (http://hilo.pmel.noaa.gov/bering/mdb/): The mission of the Bering Sea Ecosystem Biophysical Metadatabase Project is to locate, assemble and disseminate an inventory of the extensive biological and physical data collected on the subarctic, semi-enclosed Bering Sea ecosystem. Bering Sea is defined by the coasts of Russia, Alaska and the Aleutian Island chain in North Pacific Ocean and maintains a wide variety of fish, shellfish, seabirds and marine mammals.
- Bernegat Bay (http://www.crssa.rutgers.edu/projects/runj/bbay.html): Bernegat Bay Projects include a series of initiatives by the Centre for Remote Sensing and Spatial Analysis at Rutgers University for the application of geospatial technology for landscape and watershed ecological analysis for the Barnegat Bay watershed region in New Jersey (Lathrop *et al.* 2000).
- Black Sea (http://www.blackseaweb.net/): Detailed discussions among marine environmental scientists from Romania, the Russian Federation and Ukraine resulted in the opinion that data availability as well as data access and integrated marine environmental management of the Black Sea Region may be improved by application of innovative practice tools and management systems from EC countries. The Black Sea Web on the Internet provides such management tool and helps the decision makers in their tasks to implement an adequate integrated marine environmental management of the Black Sea Region. The project is a joint research between local partners from Ukraine, the Russian Federation and Romania and EC partners from the Netherlands and Denmark, focusing on strengthening of environmental management in the region (UN 1997; Belokopytov 1998).
- Chesapeake Bay (http://www.chesapeakebay.net/cims/): The Chesapeake Information Management System (CIMS) is an organised, distributed library of information and software tools designed to increase basin wide public access to Chesapeake Bay information. It includes the Bay Atlas, an interactive GIS mapping application as well as a series of other stand-alone and Internet applications.
- Glacier Bay (http://www.inforain.org/alaska/glabaycd/): The Glacier Bay Ecosystem GIS contains more than 100 different spatial data sets covering the greater Glacier Bay, located in Alaska area, stretching from Yakutat in the north to south of Admiralty Island. The system is provided as an online application as well as on CDROM.

- Massachusetts Bay (http://massbay.mit.edu/): The Massachusetts Bay Information Server is an initiative by the Sea Grant Marine Centre for Coastal Resources at the Massachusetts Institute of Technology providing research output, monitoring and GI for Massachusetts Bay.
- Massachusetts (http://www.state.ma.us/mgis/massgis.htm): The Massachusetts Geographic Information System (MassGIS) is a comprehensive, statewide database of spatial information for environmental planning and management. MassGIS is the official state agency assigned to the collection, storage and dissemination of geographic data. Part of MassGIS includes several coastal and marine features, such as coastline, fish trap locations, corridors of anadromous fish, designated shellfish growing areas, lobster harvest zones, tidal restrictions, bathymetry for the Gulf of Maine, state designated barrier beaches, federal and state marine sanctuaries and salt marsh restoration sites.
- North Sea (http://www.umweltprogramme.de/sustainabilitycentres/geograph/): The North Sea GIS is a component of the project 'Sustainability Centres in the North Sea Region'. The system is a GRASS-based GIS operating from the German Headquarters of GRASS GIS in Hannover. The GIS is available for use by planners, engineers, architects, university staff and non-governmental organisations from across the North Sea Region.
- Penobscot Bay (http://www.csc.noaa.gov/id/text/pbgis.html and http://www. penbay.net/mapping.htm): The Penobscot Bay GIS is an effort by NOAA's Coastal Services Centre to provide technical support to the National Environmental Satellite, Data, and Information Service (NESDIS) and the Island Institute of Maine for the development and maintenance of a regional GIS decision support system for Penobscot Bay in the Gulf of Maine.
- San Francisco Bay (http://www.regis.berkeley.edu/baydelta.html): The San Francisco Bay, Sacramento/San Joaquin Delta GIS is a monitoring effort by the University of California's Centre for Environmental Design Research (CEDR) aiming to the completion of a series of research projects to begin the formation of an Environmental Framework Plan for the 12 County Bay/Delta region.
- The San Francisco Bay Area EcoAtlas (http://www.sfei.org/ecoatlas/): The San Francisco Bay Area EcoAtlas is a GIS atlas developed by the San Francisco Estuary Institute in order to respond to the many public requests for paper maps and digital files for the Bay Area.
- PAGIS (http://www.csc.noaa.gov/pagis/): The Protected Areas Geographic Information System Project is a US National Ocean Service initiative to develop fully integrated GIS, spatial data management and Internet capabilities at all National Estuarine Research Reserves and National Marine Sanctuaries in the United States.
- UK-CMC (http://195.224.53.130/ukcoastmap/). The UK Coastal Map Creator is an Internet-based interactive GIS application, which provides users and managers of the UK coast with online access to geospatial data and map creation for the UK coastal zone.

Many are the interactive GIS data distribution systems available on the Internet, which mainly provide georeferenced data, based on users' selections. The Live Access to Climate Data server is a product of the 'Thermal Modelling and Analysis Project' (TMAP) at NOAA's Pacific Marine Environmental Laboratory

(PMEL). This Internet server uses an analysis tool, FERRET, to provide access to gridded environmental datasets. It enables the user to interact with a remote UNIX server and create a variety of visualisations of certain climatologies (http://ferret.wrc.noaa.gov/fbin/-climate_server). NOAA's Real Time TAO Buoy Data Display allows access to real-time data received daily via satellite from the Tropical Pacific Ocean buoy array (http://www.pmel.noaa.gov/togatao/realtime.html). The Interactive Marine Buoy Observations of the US National Weather Service allows users to select highlighted regions and receive weather and sea state data being reported by automated marine stations (http://www.nws.fsu.edu/buoy/). The Hellenic Centre for Marine Research provides buoy data and forecast animations generated by Poseidon, the oceanographic buoy array in SE Mediterranean (http://www.poseidon.ncmr.gr/). The Java enhanced Interface to EPIC Oceanographic Profiles (NOAA) is a data selection applet, which provides an interface to online databases on this server (http://www.epic.noaa.gov/epic/edsel/edsel.html). EPIC was created by PMEL to manage the large numbers of hydrographic and time series *in situ* datasets collected as part of NOAA's oceanographic and climate study programmes. The Australian Coastal Atlas (ACA, http://www.environment.gov.au/marine/coastal_atlas/) is a national network of marine and coastal agencies that are linked as Internet nodes using a variety of interactive mapping tools to provide layers of information about the Australian coastal environment including water quality, climate and fisheries. The ACA was first established in 1995 to support a related initiative, the Australian Spatial Data Directory, which is a directory of data available around Australia (ASDD, http://www.environment.gov.au/net/asdd/). The US Naval Oceanographic Office (NAVOCEANO) developed a worldwide Virtual GIS Information Server (http://www.navo.navy.mil/navdriver.html), which is designed to give users access to a list of satellite and buoy products for a given geographic region. The product list is filtered at users' selection of a geographic area of interest that results in a list of the most currently available products or information. The USGS Western Region Geologic Information provides data and images from the Pacific Sea Floor Mapping project (http://wrgis.wr.usgs.gov/dds/dds-55/pacmaps/site.htm). Sonar data (bathymetry and reflected energy off the seafloor) are provided as GIS ready coverages and full resolution GIS images.

The Land Use Coordination Office's Geographic Information Service (LUCO GIS) is custodian of several GIS datasets pertaining to land use planning and marine resource mapping in British Columbia (BC), Canada (http://www.gis.luco.gov.bc.ca/). LUCO GIS provides detailed mapping information for BC marine resources (scale 1:40 000) through a variety of inventories and products, such as the Coastal Resource Inventory (a biophysical inventory of coastlines using a combination of helicopter video and field sampling), Marine Ecological Mapping (ecoprovinces, ecoregions and ecosections), Marine Protected Areas and Coastal Resource Atlases (shoreline sensitivity to oiling, commercial herring and salmon fisheries and recreational fisheries). The Ocean Biogeographical Information System (OBIS, Colour plate 1), an international research programme assessing and explaining the diversity, distribution and abundance of marine organisms throughout the world's oceans, is envisioned to be a distributed network of marine biological and environmental data for use in examining the changes in diversity, distribution and abundance of organisms over time and space. OBIS will contain an analysis tool for arriving at multidisciplinary answers to important scientific and societal questions. The system will bridge the gap between georeferenced data that are stored in traditional non-standard or standard science formats and geospatial data that are stored in a way designed to be accessed and analysed using a GIS (http://core.cast.msstate.edu/censobis1.html).

The Mediterranean Forecasting System Pilot Project is an initiative among more than 25 institutions from Italy, France, Great Britain, Norway, Cyprus, Malta, Egypt, Spain, Greece and Israel (http://www.cineca.it/%7Emfspp000/), resulting in a forecasting system with two essential parts, an observing system and a numerical modelling/data assimilation component that can use the past observational information to optimally initialise the forecast. GOOS (Global Ocean Observing System) are internationally organised systems for the gathering, coordination, quality control and distribution of many types of marine and oceanographic data and derived products of common worldwide importance and utility (http://ioc.unesco.org/goos/). GOOS are supplemented through various support organisations by financial, manpower and in kind contributions. These international organisations include IOC (Intergovernmental Oceanographic Commission of UNESCO), WMO (UN World Meteorological Organisation), UNEP (UN Environment Programme) and ICSU (International Council of Scientific Unions). The scientific design of GOOS includes several modules (climate, health of the oceans, living marine resources, coastal and services modules). Several of these observing systems have already been implemented worldwide: EuroGOOS, MedGOOS (Mediterranean), GOOSAfrica, NEAR/GOOS (NE Asian), PacificGOOS, IOCARIBEGOOS (Caribbean) and SEAGOOS (SE Asia). Johannessen *et al.* (1997) reviewed the possible applications of earth observation data within EuroGOOS, underlining the need of satellite data integration for monitoring purposes of oceanographic processes.

During the last decade, GIS technology is widely used in all four major branches of ocean science: (1) geology; (2) chemistry; (3) physics; and (4) biology. The spatial attribute of oceanographic data makes GIS a suitable data management and analysis tool for a wide range of oceanographic applications covering wetland and coastal environments as well as open ocean surface and deep regions. Developments of oceanographic GIS as database management tools provide a highly suitable platform for the appropriate storage of the various geolocated ocean data. The placement of data in such spatially referenced database systems facilitates ocean data analysis, integration and visualisation. The mapping of ocean bottom provides new information on deep ocean environments, mapping of benthic habitats provides new approaches to their management, mapping of underwater landslides provides new information for marine geomorphology and mapping of coastal environments provides the information needed for their sustainable management. In addition, the mapping and monitoring of oceanographic processes under GIS provide an analytical tool for the establishment of seasonality on the physical dynamics of marine environment. In connection with the Internet, oceanographic GIS provide common information to different management authorities, information-based management scenarios, online mapping tools for the generation of spatiotemporal data maps, online management tools for specific marine areas, easy communication of results from similar studies and easy information exchange on the development of methodological aspects in oceanographic GIS. In addition, online connection of oceanographic GIS with ocean forecasting systems, although a practise not highly developed yet, will further provide tools for the prediction of dynamics of oceanographic processes and natural hazards.

## 2.5 OCEANOGRAPHIC DATA SAMPLING METHODS

Scientists are using world leading rapid assessment techniques to survey and map the world's marine environments. Investment in technology to support marine sciences generates baseline data on the physical, chemical and biological environment. This knowledge is integral to the development of increasingly sophisticated decision support systems used in management of the marine environment and to understanding short, medium and long-term variations in oceans, fisheries and climate. Oceans are continually monitored by a number of different sensors on board satellites, airplanes and vessels as well as attached to moored and drifting buoy systems.

Satellites orbiting thousands of kilometres from earth, carry a variety of sensors that can generate terabytes of data each day on the state of the oceans from their surface temperature and surface height indicating the presence of ocean currents and eddies to surface winds and waves, tidal variations, sea ice and the ocean colour, which varies with nutrient content. These optical sensors may gather data 'passively' by recording the light reflected by surface features or 'actively' by emitting a beam of light and measuring its response. Once captured, images of the area of interest are passed on to analysts who interpret the data, extract information and use it to answer questions. By using these sensors, scientists and researchers can remotely gather more information than they can physically sense. The use of RS, which is the gathering of information over an area using a device not in direct physical contact with the area being studied, made possible the frequent collection and analysis of information about large oceanic areas.

Marine EO from satellites is particularly suitable for sampling and monitoring large areas. RS data on ocean surface recorded by EO satellites provide timely and geographically specific information essential to early warning on natural disasters, sustainable development and management of natural resources, monitoring coastal zones and environmental protection. The EO capacities have been greatly enlarged since the introduction of radar imaging systems to EO from satellites. The synthetic aperture radar (SAR) RS systems are complementary to the optical RS systems, because they can record clear images through clouds, rain or snow, as well as at night. This all weather, day and night imaging capacity of SAR RS systems is particularly important for data collection in areas having high frequency of cloud cover (e.g. equatorial regions and mid/high latitude western margins) and in emergency situations, such as during floods, as well as for routine, operational applications where reliability of frequent image data acquisition is critical (e.g. ice monitoring).

The growing number of EO satellites in orbit and the continuing improvements of their RS systems (sensors) provide mankind with unprecedented global capacity for systematic monitoring of the oceans. Along with the beginning of the third millennium, globalisation is becoming an increasingly dominant factor for the well-being of countries. Large companies have spearheaded this trend in the economic sector and established multinational corporations to be able to better serve the global markets and thus to survive. Similarly, planning of sustainable development strategies has to be coordinated with neighbouring countries and harmonised at a global level in order to succeed. The EO, being inherently global, provides the most important source of geospatial information for the implementation of large and often complicated development strategies.

The major EO products about the ocean's surface include temperature distribution, CHL concentration and altimetry information. The Coastal Zone Colour Scanner (CZCS), which was active from 1978 to 1986, proved that discoloration of the water caused by chlorophyll present in phytoplankton is detectable by satellite. The substantial CZCS and Ocean Colour and Temperature Scanner (OCTS: September 1996 to June 1997) archives serve as data sources for historical information on phytoplankton concentrations worldwide. The ocean colour satellite, Sea viewing Wide Field of view Sensor (SeaWiFS), launched in August 1997, is essential for current and future forecasting capabilities providing CHL concentration, as a phytoplankton biomass parameter and attenuation coefficients, as indicators of water clarity. Derived marine and atmospheric products include absorbing aerosol index, CHL, cloud (fraction, optical depth and top height), calcium carbonate, coccolithophorids, dissolved detritus absorption, harmful algal blooms, photosynthetically active radiation (PAR), particulate backscatter coefficient, suspended sediments, smoke index and *Trichodesmium* distribution. The SeaWiFS imagery products are provided as daily, weekly, rolling 32-day composites and monthly climatologies in full and reduced resolutions. The Advanced Very High Resolution Radiometer (AVHRR) has been used to identify physical oceanographic features by detecting differences in surface water temperature. For example, intrusions of warm water onto continental shelves and near shore upwelling can be detected in AVHRR imagery. The combined use of SeaWiFS and AVHRR data provides the ability to identify features that concentrate and transport phytoplankton as well as to map the spatial extent of the main oceanographic processes.

Imagery from microwave altimeters on board TOPEX/Poseidon and European Remote Sensing (ERS-2) satellites can be used to identify physical oceanographic features by detecting differences in surface water height. The ability to identify features such as geostrophic currents that concentrate and transport phytoplankton can be used as an early warning of conditions conducive to fish aggregation. Also, information on sea surface elevation is important for the prediction of tides and storm surges. Ocean currents and fronts can be identified by SAR that detects surface roughness. This imagery allows identification of physical oceanographic features that concentrate and transport phytoplankton. These microwave sensors are available on ERS-2 and the Canadian RADARSAT satellite.

Wind fields are most useful in determining the direction and velocity of phytoplankton transport. Large-scale wind patterns can provide information critical to determining the movements of surface ocean features. Winds are a major factor in the initiation of the major oceanographic processes by mixing the water column, thereby disrupting the natural environment. Data are available from the Advanced Microwave Instrument (AMI) aboard ERS-2 and from the Special Sensor Microwave/Imager (SSM/I).

Johannessen *et al.* (2000) provided a review of available satellite earth observation data and their potential and suitability for use in operational oceanography. They underlined that oceanographic satellite data must be used in synergy with various data assimilation methods (e.g. oceanographic models and GIS) in order to fully benefit from their use. Table 2.1 lists the major EO marine satellites that daily sense the marine environment from space.

**Table 2.1.** Various satellite sensors from which acquired data are most commonly used in marine GIS applications.

| SENSOR | SATELLITE | INTERNET SOURCE |
| --- | --- | --- |
| Active Microwave Instrument (AMI) | ERS-2 | http://earth.esa.int/ipgami |
| Advanced Very High Resolution Radiometer (AVHRR) | POES | http://edcdaac.usgs.gov/1KM/avhrr.sensor.html |
| Along Track Scanning Radiometer (ATSR) | ERS-2 | http://earth.esa.int/ers/ |
| Dual frequency Altimeter (ALT) | TOPEX/Poseidon | http://topex-www.jpl.nasa.gov/ |
| Moderate resolution Imaging Spectroradiometer (MODIS) | Terra | http://modis.gsfc.nasa.gov/MODIS/ |
| Modular Optoelectronic Scanner (MOS) | IRS-P3 | http://www.ba.dlr.de/NEWS/ws5/mos_home.html |
| Multiangle Imaging Spectroradiometer (MISR) | Terra | http://www-misr.jpl.nasa.gov/ |
| Ocean Colour Imager (OCI) | ROCSAT-1 | http://rocsat1.oci.ntou.edu.tw/en/oci/ |
| Ocean Colour Monitor (OCM) | IRS-P4 | http://www.isro.org/programmes.htm |
| Ocean Scanning Multispectral Imager (OSMI) | KOMPSAT | http://kompsat.kari.re.kr/english/ |
| Sea viewing Wide Field of view Sensor (SeaWiFS) | OrbView-2 | http://seawifs.gsfc.nasa.gov/SEAWIFS.html |
| Seawinds | Quickscat | http://winds.jpl.nasa.gov/ |
| Solid-State Altimeter (SSALT) | TOPEX/Poseidon | http://www.jason.oceanobs.com/html/portail/general/welcome_uk.php3 |
| Synthetic Aperture Radar (SAR) | ERS-2, RADARSAT-1 | http://radarsat.mda.ca/ |

Sensing platforms on board aircrafts is another means of RS. Aircraft monitoring may be most useful to detect large-scale features that occur close to the shore, where satellite data are often invalid because of interference by continental aerosols and terrestrial features. In addition, aircraft-based videography and photography have huge advantages over RS in terms of spatial resolution, purpose of sampling and selectivity of flight paths and times, although data is collected only for relatively smaller areas. Because they are flown at low-altitude, aircraft-based instruments are relatively unaffected by cloud cover. Aircrafts can be equipped with several instruments for simultaneous measurement of several variables. If the systems are automated and user-friendly, they could be mounted to 'Aircraft of Opportunity' such as the US Coast Guard helicopters that routinely fly over and monitor the coastal regions of the Gulf of Mexico. Ocean colour can be measured using passive optical systems such as the Ocean Data Acquisition System (ODAS) and SeaWiFS Aircraft Simulator (SAS) that measure radiance at selected visible wavelengths. Examples are the Airborne Visible Infra Red Imaging Spectrometer (AVIRIS) and the Compact Airborne Spectrographic Imager (CASI). An alternative to ocean colour passive measurements, active laser fluorescence (Light Detection and Ranging, LIDAR) has been used to detect phytoplankton by measuring CHL pigment fluorescence. LIDAR can be modified to detect fluorescence of coloured dissolved organic material and phycoerythrin containing plankton such as cyanobacteria, which could prove useful in the detection of *Trichodesmium* spp. Also, several systems of LIDAR bathymetry have now reached operational maturity and are widely used in hydrographic surveys. Smith and West (1999) discussed the characteristics of airborne LIDAR bathymetry and how it can be best employed to integrate with other survey systems. Finally, the Advanced Airborne Hyperspectral Imaging System (AAHIS) is a high-resolution spectrometer that is widely used for the development of habitat sensitivity index mapping. The sensor is optimised for littoral region RS for a variety of civilian and defense applications including ecosystem surveying and inventory, detection and monitoring of environmental pollution, infrastructure mapping and surveillance (Holasek *et al.* 1997).

On the other hand, ocean engineering centres design, develop and maintain instruments that can sense beneath the ocean surface and provide scientists with environmental information about water properties and characteristics of marine life. For many ocean observations, instrumentation has been developed and refined over many years of research into reliable ready to use instruments. These instruments may be found on any ocean going research vessel. Some of these instruments are permanently mounded to commercial ships and provide oceanographic data through various 'Ships of Opportunity' programmes.

The main oceanographic instruments are the Conductivity/Temperature/Depth (CTD) probe, the Acoustic Doppler Current Profiler (ADCP), and the Expendable BathyThermograph (XBT). The CTD probe is a mainstay of ocean research, providing details of temperature, salinity and nutrients, dissolved oxygen and chlorofluarocarbons to depths of 6000 m. This electronic measuring system is used from research vessels that remain at a station or geographic location while the CTD is lowered to sample the water at different depths. The ADCP is fixed to the hull of a research ship sending signals to a depth of 300 m to measure current speeds at different depths. The XBT takes continuous measurements of temperature and salinity to a depth of 1200 m. Both research vessels and commercial ships that participate in various 'Ships of Opportunity' programmes throughout the globe use XBT. Also, the Continuous Plankton Recorder (CPR), developed at Plymouth Marine Laboratory in United Kingdom, is a low-cost plankton sampler, which is designed for both towed and moored deployment and is increasingly being used worldwide in oceanographic monitoring programmes.

Another important source of oceanographic data is the buoy. Buoys can be kept on site as moored systems or left drifting in the coastal and open ocean for extended periods. Buoy instruments provide vertical profile data that are not available readily from any other RS platform. Buoys have the potential of being equipped with several instruments for simultaneous measurement of several variables, such as wind speed and direction, air temperature and atmospheric pressure, subsurface water temperature and distribution of phytoplankton populations (by measuring irradiance and absorption coefficients), salinity, currents, depth, etc. For example, in the past 10 years, CSIRO (Commonwealth Scientific and Industrial Research Organisation, Australia) has built and deployed hundreds of moored instrument arrays. In the World Ocean Circulation Experiment from 1994 to 1996, moored instruments were deployed in the Indian, Pacific and Southern oceans to obtain temperature and salinity data, which also revealed the size of ocean currents, their speed and direction. The largest permanent buoy array today is the TAO/TRITON array (Tropical Atmosphere Ocean), which consists of approximately 70 moorings appropriately placed throughout the Tropical Pacific Ocean. The purpose of this array is the telemetry of oceanographic and meteorological data to shore in real-time via the Argos satellite system. The array is designed to improve detection, understanding and prediction of El Nino. It is a major component of the El Nino/Southern Oscillation (ENSO) Observing System, the Global Climate Observing System (GCOS) and the Global Ocean Observing System (GOOS). It is supported by US (National Oceanic and Atmospheric Administration) and Japan (Japan Marine Science and Technology Centre) with contributions from France (Institut de Recherche pour le Developement).

The first generation of torpedo shaped, satellite tracked surface drifter buoys began tracking pathways of the major ocean currents around the globe since the 1970s. The latest generation of drifting profilers can sink to preprogrammed depths of 2000 m and obtain temperature and salinity profiles. Periodically, they surface automatically and upload their data to a passing satellite. Each profiler has a life of 2 years and once deployed, the profilers cannot be recovered. Their main application is ocean and climate research in coastal and open waters. Their data are used for prediction of regional and worldwide ocean and climate trends. Many of the autonomous floats now carry CTD sensors and transmit profiles of temperature and salinity each time they surface. Such floats are the mechanism by which a planned global array of 3000 floats (the Argo array) will provide upper ocean temperature and salinity information from world oceans into the twenty-first century (Argo Science Team 1998).

The amount of oceanographic information derived from RS data and other research instruments has significantly increased during the last decades. The range of their applications broadened, especially when the optical RS and SAR RS data, as well as relevant geospatial data from other research instruments are integrated in GIS and jointly analysed. While there are many applications for which RS datasets from a single RS system will be adequate, the possibility of selective integration of diverse RS and other geospatial datasets in GIS and their joint analysis is a major recent advancement. It will result in greater impact of EO on sustainable management of natural resources and environmental protection and in enhanced social benefits.

## 2.6 OCEANOGRAPHIC DATA SOURCES AND GIS DATABASES

Today, most environmental agencies and administrations provide their raw and processed data through online, publicly available data archives. The list of oceanographic data sources that is provided here is by no means complete. The list includes Internet addresses for main oceanographic data, such as SST, sea surface CHL concentration, sea surface altimetry products, and 3D time series and climatologic datasets (Table 2.2). Depending on the nature of the raw data, data products are provided mainly in various image formats and time series spreadsheets, which makes it relatively easy for GIS input while several datasets are provided in GIS ready formats.

Many international and national organisations maintain digital data archives and freely offer datasets for research and educational purposes. One of the most comprehensive data management and distribution networks in the world is the World Data Centre System. World Data Centres exist in all continents and have computer facilities using electronic networks to meet user requests, exchange data catalogue information and transfer data (http://www.ngdc.noaa.gov/wdc/wdcmain.html). In the United States, a nationwide network of distributed data archive centres (DAAC) is established and distributes data regarding the physical and biological state of the oceans and offers educational CDROMs and various guide documents. In Europe, data holding centres in Germany, France, United Kingdom, Belgium and Italy distribute various oceanographic data for the Mediterranean, Black Sea and the Atlantic Ocean. In Africa, Asia, and Australia similar centres offer their data freely on the Internet. Worldwide, many centres provide real-time oceanographic data. Depending on the spatiotemporal extent of data coverage, the spatial and temporal resolutions of the available data differ. However, based on these geolocated data, GIS analysis of the major oceanographic processes becomes possible.

Data from three sensors are considered adequate for basic sea surface analysis. The Sea viewing Wide Field of view Sensor (SeaWiFS) provides, among a series of products, worldwide concentration of sea surface chlorophyll. The Advanced Very High Resolution Radiometer (AVHRR) provides worldwide distribution of SST. The TOPEX/Poseidon provides a series of altimetry products, which include currents, sea surface height, wind and waves, sea-level anomaly and bathymetry. Three-dimensional climatologic datasets of various parameters provide adequate information about the state of the water column. The Naval Oceanographic Office Data Warehouse Model and many buoy data providing centres offer 3D profiles for the major oceanic areas. Global measurements of ocean temperature and salinity are accessible to users via the Global Temperature/Salinity (T/S) Profile Programme (GTSPP). A cooperative international project, GTSPP maintains a global ocean T/S resource with both real-time data transmitted over the Global Telecommunications System (GTS) and delayed mode data received by NOAA National Oceanographic Data Centre (NODC). Countries contributing to the project include Australia, Canada, France, Germany, Japan, Russia and US (IOC 1996a,b). The Mediterranean Oceanic DataBase (MODB) provides 3D climatological profiles for the whole Mediter-ranean basin (Brasseur *et al.* 1996).

The German Federal Maritime and Hydrographic Agency (BSH, Bundesamt fur Seeschiffahrt und Hydrographie) provides weekly products of SST for the North and Baltic seas (http://www.bsh.de/Oceanography/Climate/Actual.htm). BSH has published weekly composite SST analyses since 1968. Until November 1994 the mapping technique consisted in plotting ship and station data onto maps but since December 1994 AVHRR SST data are used. The German Aerospace Centre (DLR, Deutschen Zentrum fur Luft und Raumfahrt) provides daily, weekly and monthly products of SST for the Mediterranean and Black Sea and NE and SE Atlantic. DLR disseminates these image products through a network interface called Graphical Intelligent Satellite Data Information System (GISIS, http://isis.dlr.de/). NASA's SeaWiFS Project at the Goddard Space Flight Centre disseminates various worldwide image products derived from the SeaWiFS sensor (http://seawifs.gsfc.nasa.gov/SEAWIFS.html). NASA's TOPEX/Poseidon Project at the Jet Propulsion Laboratory disseminates worldwide TOPEX products (http://topex-www.jpl.nasa.gov/). Data from the TAO/TRITON array in the Tropical Pacific are available through the TAO/TRITON Data Delivery system (http://www.pmel.noaa.gov/tao/). The US Naval Oceanographic Office (http://www.navo.navy.mil/navdriver.html) developed a worldwide marine database, the Virtual GIS, which is designed to give the user immediate access to a list of all the currently available products for a given geographic region. All of the listings provided are dynamically generated and the resulting list contains the most currently available products or information. Plewe (1997) discussed the subject of using GIS as online data distribution tool and listed many sites that are distributing data.

**Table 2.2.** Major oceanographic data providers on the Internet.

| ORGANISATION | WEBSITE URL | NOTES |
|---|---|---|
| NAVOCEANO | http://128.160.23.42/dbdbv/dbdbv.html | Worldwide bathymetry |
| | http://128.160.23.42/gdemv/gdemv.html | Worldwide CTD profiles |
| AVISO Altimetry | http://alti.cnes.fr/ | Worldwide altimetry products |
| USGS/Sediment Spatial Data | http://atlantic.er.usgs.gov/habitat/wflopen2/htm/aview.htm | Various US sonar generated data |
| Aerial photography for Puerto Rico | http://biogeo.nos.noaa.gov/data/ | Aerial photography for US Virgin Islands and Estuarine Living Marine Resources for the Gulf of Mexico. |
| NOAA/Office of Ocean and Coastal Resource Management | http://cammp.nos.noaa.gov/cammp/ | Database of coastal zone management programmes in US |
| US/ACE, The Coastal Data Information Program | http://cdip.ucsd.edu/ | Wave, wind and temperature data from oceanographic stations in US |
| CEONET, Canada | http://ceonet.ccrs.nrcan.gc.ca/cs/en/index.html | Worldwide database of various oceanographic and other data |
| California Resources Agency | http://ceres.ca.gov/ocean/infotype/skel/maps.phtml | Various oceanographic data for the western US |
| USGS/East Gulf of Mexico Satellite Imagery | http://coastal.er.usgs.gov/east_gulf/ | SST, reflectance and altimetry images for the Gulf of Mexico |
| US Geological Survey Geological Division | http://crusty.er.usgs.gov/coast/getcoast.html http://rimmer.ngdc.noaa.gov/coast/ | Worldwide coastline |
| Coastal Marine Profile and Time Series Browser | http://crusty.er.usgs.gov/epic/ | Currents, temperature, salinity, light attenuation, subsurface pressure for US |
| NOAA/NASA Pathfinder Program | http://daac.gsfc.nasa.gov/CAMPAIGN_DO CS/LAND_BIO/path_sites.html | Various satellite data |
| NASA/Data Assimilation Office | http://dao.gsfc.nasa.gov/ | TMI and SSM/I Data source |

| | | |
|---|---|---|
| University of Rhode Island/SST Satellite Image Archive | http://dcz.gso.uri.edu/avhrr-archive/archive.html | AVHRR SST images for NW Atlantic |
| CEOS, Committee on Earth Observation Satellites | http://dial.eoc.nasda.go.jp/ | OCTS (Ocean Color and Thermal Scanner) datasets |
| GEOLIST/Data Sources | http://dspace.dial.pipex.com/geolist/htmdbase/data.htm | List of various worldwide data holding centres |
| National Center for Atmospheric Research | http://dss.ucar.edu/catalogs/free.html | Worldwide oceanographic and atmospheric datasets |
| USGS EROS Data Center | http://edcdaac.usgs.gov/ | Satellite imagery (Landsat 7 ETM+), Terra (ASTER and MODIS), Landsat Pathfinder, AVHRR, Elevation (Global 30 Arc Second Elevation Data Set), and RADAR (SIR-C) data |
| EROS Data Center | http://edcwww.cr.usgs.gov/eros-home.html | Aerial photography for the US and worldwide satellite images |
| Ocean Remote Sensing, Johns Hopkins University | http://fermi.jhuapl.edu/ | AVHRR and SAR imagery for the US |
| NOAA/FERRET | http://ferret.wrc.noaa.gov/las/main.pl | Various worldwide datasets |
| SADCO/South African Data Center for Oceanography | http://fred.csir.co.za/ematek/sadco/sadco.html | Station profile data of temperature, salinity, oxygen, nutrients, wind, swell for oceans around central and south Africa |
| Global Change Master Directory | http://gcmd.gsfc.nasa.gov/ | Various worldwide oceanographic datasets |
| GeoGratis, Canada | http://geogratis.cgdi.gc.ca/frames.html | Various satellite data for Canada |
| Global Change Data and Information System | http://globalchange.gov/ | Getaway to global change data |
| U.S. GLOBEC Data System | http://globec.whoi.edu/jg/newdir | Various oceanographic data for Georges Bank, NE Pacific and Southern Ocean |
| International Arctic Buoy Programme | http://iabp.apl.washington.edu/ | Various buoy data for the Arctic basin |
| NOAA Laboratory for Satellite Altimetry, Sea Floor Topography | http://ibis.grdl.noaa.gov/cgi-bin/bathy/bathD.pl | Global sea floor topography based on satellite altimetry and ship depth soundings |
| OceanPortal | http://ioc.unesco.org/oceanportal/welcomeframe.htm http://ioc.unesco.org/iode/home.htm | Directory of ocean data and information related websites |
| International Research Institute for Climate Prediction | http://iri.ldeo.columbia.edu/ | Worldwide meteorological station data |
| German Remote Sensing Data Center (DFD) | http://isis.dlr.de/ | Marine satellite data for Eastern Atlantic and the Mediterranean |
| Current Meter Data Assembly Center | http://kepler.oce.orst.edu/ | Worldwide oceanographic buoy data |

| NOAA/The Satellite Active Archive | http://las.saa.noaa.gov/ | Worldwide marine and atmospheric satellite images |
|---|---|---|
| NASA/JPL/LinkWinds | http://linkwinds.jpl.nasa.gov/ | Software for processing wind data |
| Rutgers University, Institute of Marine and Coastal Sciences | http://marine.rutgers.edu/mrs/data.html http://marine.rutgers.edu/mrs/sat.data2.html | Satellite, weather and radar images for NW Atlantic |
| Space Application Institute, Marine Environment Unit | http://me-www.jrc.it/ http://me-www.jrc.it/OCEAN/ocean.html | Marine satellite images for European Seas and North Indian Ocean |
| Mediterranean Oceanic Data Base | http://modb.oce.ulg.ac.be/ | Profile data and bathymetry for the Mediterranean |
| NSIPP TOPEX DATA ARCHIVE | http://mohawk.gsfc.nasa.gov/topexdat/ | Worldwide TOPEX/Poseidon altimetry products |
| NASA's Distributed Active Archive Centers | http://nasadaacs.eos.nasa.gov/ | List of NASA's Distributed Active Archive Centers (DAACs) |
| NEMO/Oceanographic Data Server | http://nemo.ucsd.edu/ | Station, bathymetry and wind datasets |
| World Ocean Circulation Experiment Data Information | http://oceanic.cms.udel.edu/woce/ | Various satellite and surveyed worldwide data |
| NOAA/Satellite Imagery on the Internet | http://orbit35i.nesdis.noaa.gov/arad/fpdt/nwasat.html | List of sites providing weather satellite imagery |
| NASA Jet Propulsion Laboratory Physical Oceanography Distributed Active Archive Center | http://podaac.jpl.nasa.gov/ http://podaac-www.jpl.nasa.gov/othersources.html | Wind and SST climatologies |
| World Ocean Circulation Experiment Satellite Data CDROM, JPL/NASA | http://podaac.jpl.nasa.gov/cdrom/woce/ | Online CDROM with altimetry and AVHRR products |
| NOAA/NASA AVHRR Oceans Pathfinder | http://podaac-www.jpl.nasa.gov/sst/ | Worldwide SST images |
| NOAA/Pacific Fisheries Environmental Laboratory | http://salmonid.pfeg.noaa.gov/las.html | Worldwide collection of oceanographic products and upwelling indices |
| National Data Buoy Center | http://seaboard.ndbc.noaa.gov/ | Meteorological and oceanographic buoy data for US |
| NOAA/National Ocean Service | http://seaserver.nos.noaa.gov/projects/shoreline/shoreline.html | Coastline for the US |
| Sea Wide Field-of-view Sensor Project | http://seawifs.gsfc.nasa.gov/SEAWIFS.html | Worldwide SeaWiFS data products |
| NOAA/CoastWatch | http://sgiot2.wwb.noaa.gov/COASTWATCII/ | Satellite data for Western Atlantic |
| NASA/Terra | http://terra.nasa.gov/ | List of data holding sites |
| Terraserver | http://terraserver.microsoft.com/default.asp | Aerial and satellite photography for the US |
| Joint Archive for Sea Level-Hawaii | http://uhslc.soest.hawaii.edu/uhslc/jasl.html | Sea-level data |

| | | |
|---|---|---|
| National Geophysical Data Center (NGDC), U.S.A. | http://web.ngdc.noaa.gov/ | Satellite images and marine geology datasets |
| Maritime Claims Manual | http://web7.whs.osd.mil/html/20051m.htm | Maritime claims reference manual for almost all coastal countries |
| JPL-NASA | http://winds.jpl.nasa.gov/ | Scatterometry (wind) data |
| USGS/Mapping Surveys | http://woodshole.er.usgs.gov/operations/sfmapping/surveys.htm | Sea floor mapping data from USGS surveys |
| Australian Oceanographic Data Center | http://www.AODC.gov.au/AODC.html | Various gridded oceanographic datasets around Australia |
| Drifting Buoy Data Assembly Center | http://www.aoml.noaa.gov/phod/dac/dac.html | Worldwide drifter buoy data |
| Black Sea Web | http://www.blackseaweb.net/ | Various oceanographic data for the Black Sea |
| British Oceanographic Data Center | http://www.bodc.ac.uk/ | Various worldwide datasets and data visualisation software |
| Bureau of Meteorology, Australia | http://www.bom.gov.au/ | Various meteorological and oceanographic information for Australia and Antarctic |
| Deutsches Ozeanographisches Datenzentrum | http://www.bsh.de/3172.htm | SST images for the North Sea |
| German Oceanographic Data Center | http://www.bsh.de/Oceanography/DOD/DOD.htm | Various surveyed data and list of data holding centres in Germany |
| Comprehensive Ocean-Atmosphere Dataset | http://www.cdc.noaa.gov/coads/ | Worldwide images of SST, air temperature, wind, sea-level pressure, cloudiness and relative humidity |
| Barnegat Bay projects/ Rutgers University | http://www.crssa.rutgers.edu/projects/runj/bbay.html | Various data for Barnegat Bay, US |
| NOAA/Coastal Services Center | http://www.csc.noaa.gov/products/shorelines/ | Shoreline for the US |
| NOAA/Coastal Shoreline Website | http://www.csc.noaa.gov/shoreline/ | Worldwide shoreline and LIDAR data |
| Atlantic Coastal Database Directory | http://www.dal.ca/aczisc/acdd | List of more than 500 databases related to integrated management of the coastal zone of Atlantic Canada. |
| German Remote Sensing Data Center (DFD) | http://www.dfd.dlr.de/ | Various satellite worldwide products |
| Natural Resources and Environmental Management, Univ. of Rhode Island | http://www.edc.uri.edu/ | Rhode Island Geographic Information System |
| NOAA/Environmental Information Services | http://www.eis.noaa.gov/esdim/ | Global change data and information service |

| | | |
|---|---|---|
| National Space Development Agency, Hatoyama Earth Observation Center, Japan | http://www.eoc.nasda.go.jp/homepage.html | Various TRMM satellite products |
| Earth Observation Research Center, Japan | http://www.eorc.nasda.go.jp/ADEOS/ | Various satellite products |
| NASDA/EORC Earth Observation Data Gallery | http://www.eorc.nasda.go.jp/EORC/Gallery/ | Worldwide satellite products |
| NOAA/Pacific Marine Environmental Laboratory | http://www.epic.noaa.gov/epic/ | Various software for online data access |
| Marine and Coastal Data Directory of Australia | http://www.erin.gov.au/ | Main oceanographic products for Australia |
| Hokkaido University | http://www.fish.hokudai.ac.jp/service/husac/sat/noaa/daily/daily_index.htm | SST for subarctic NW Pacific |
| GIS Data Depot | http://www.gisdatadepot.com/ | Various worldwide datasets |
| GRID/Geneva (UNEP) | http://www.grid.unep.ch/ | Various worldwide datasets |
| Baltic Sea Region GIS, Maps and Statistical Database | http://www.grida.no/baltic/ | Baltic Sea region datasets |
| GRID/Arendal | http://www.grida.no/db/gis/prod/html/complete.htm | Various datasets per continent |
| CERSAT/IFREMER | http://www.ifremer.fr/cersat/ http://www.ifremer.fr/cersat/ACTIVITE/E_CERACT.HTM | Various wind satellite products |
| Tropical Atlantic Pseudostress and SST Analyses | http://www.ifremer.fr/ird/WOCE/html/atlmonyr.html | Worldwide SST and pseudostress images |
| French National Oceanographic Data Center: IFREMER/SISMER | http://www.ifremer.fr/sismer/ | IFREMER marine databases |
| MEDATLAS | http://www.ifremer.fr/sismer/program/medatlas/gb/gb_medat.htm | Mediterranean hydrographic atlas |
| INFORAIN | http://www.inforain.org/dataresources/datalayers.asp | Various datasets for Western US, Alaska and Canada |
| Center of Processing and Storing the Space Information, Russia | http://www.ire.rssi.ru/cpssi/cpssi.htm | Worldwide databases of various oceanic and atmospheric parameters |
| Japan Oceanographic Data Center (JODC), Tokyo | http://www.jodc.jhd.go.jp/ | List of data holding centers |
| Lamont Doherty Earth Observatory | http://www.ldeo.columbia.edu/datarep/index.html | List of marine datasets for the US |
| Bedford Institute of Oceanography | http://www.mar.dfo-mpo.gc.ca/science/ocean/home.html | Spatiotemporal query applications to Ocean Sciences databases |
| CSIRO/Remote Sensing Project | http://www.marine.csiro.au/~lband/ http://www.marine.csiro.au/datacentre/ | SST, SAR, SeaWiFS and OCTS data for Australia |
| Irish Marine Data Center, ISMARE, Dublin | http://www.marine.ie/datacentre/ | National marine data archive for Ireland |

| | | |
|---|---|---|
| Univ. of S. Florida, College of Marine Science | http://www.marine.usf.edu/~xtuser/ | Gulf of Mexico, US East Coast and Caribbean SST images |
| MarineGIS | http://www.marinegis.com/ | List of marine data centres |
| Marine Information Service (MARIS), Netherlands | http://www.maris.nl/ | Various datasets for the European Seas |
| Marine Environmental Data Service, Canada | http://www.meds-sdmm.dfo-mpo.gc.ca/ | Various oceanographic buoy and satellite data for Canada |
| Israel Space Agency – Middle East Interactive Data Archive | http://www.nasa.proj.ac.il/ | Various meteorological and oceanographic data for Israel |
| US Naval Oceanographic Office | http://www.navo.navy.mil/navdriver.html | Worldwide 'Virtual GIS' |
| Naval European Meteorology and Oceanography Center | http://www.nemoc.navy.mil/ | Various meteorology and oceanographic products for the Mediterranean, Baltic and Black seas. |
| BAS/Antarctic Digital Database | http://www.nerc-bas.ac.uk/public/magic/add_home.html | Various data for the Antarctic |
| National Environmental Satellite, Data, and Information Service | http://www.nesdis.noaa.gov/ | US national climatic, oceanographic and geophysical data center |
| NOAA/National Geophysical Data Center | http://www.ngdc.noaa.gov/ | Bathymetry, satellite and various geophysical worldwide datasets |
| International Council of Scientific Unions | http://www.ngdc.noaa.gov/wdc/wdcmain.html | The World Data Center System |
| National Imagery and Mapping Agency | http://www.nima.mil/gns/html/ | Database of worldwide geographic feature names |
| National Oceanographic Data Center | http://www.nodc.noaa.gov/ | Worldwide chlorophyll, nutrients, ocean currents, oxygen, plankton, salinity, sea level, temperature and wave datasets |
| NOAA/Buoy Locations, Information and Recent Data | http://www.nodc.noaa.gov/BUOY/buoy.html | Worldwide buoy data |
| Global Temperature/ Salinity Profile Program Database | http://www.nodc.noaa.gov/GTSPP/gtspp-home.html | List of worldwide data CDROMs |
| National Operational Hydrologic Remote Sensing Center | http://www.nohrsc.nws.gov/ | Detailed digital coastlines for Alaska, Canada and US |
| NOAA/National Virtual Data System | http://www.nvds.noaa.gov/ | Extensive US data archived in categories: climatic, geophysical, oceanographic, radar and satellite |
| NOAA/National Weather Service | http://www.nws.noaa.gov/er/lwx/marine.htm | Satellite and buoy data for N. Atlantic |
| Pacific Marine Environmental Laboratory | http://www.pmel.noaa.gov/epic/ | Extensive surveyed oceanographic data and data visualisation tools |

| NOAA/Pacific Marine Environmental Laboratory | http://www.pmel.noaa.gov/home/data.shtml | List of various data archives |
|---|---|---|
| Tropical Atmosphere Ocean Project | http://www.pmel.noaa.gov/tao/ | Moored ocean data for Central Pacific (TAO) |
| Beaufort Sea Meteorological Monitoring and Data Synthesis Project | http://www.resdat.com/mms/ | Surveyed meteorlogical and oceanographic data for the Beaufort Sea |
| Dundee Satellite Receiving Station | http://www.sat.dundee.ac.uk/auth.html | AVHRR, MODIS and SeaWiFS image archive |
| SEA/SEARCH (EUROPE) | http://www.sea-search.net/ | Gateway to oceanographic and marine data in Europe |
| School of Ocean and Earth Science and Technology, Hawaii | http://www.soest.hawaii.edu/wessel/gshhs/gshhs.html | Global high-resolution digital shoreline |
| Distributed Oceanographic Data System | http://www.unidata.ucar.edu/packages/dods/ | List of data archives |
| Naval Research Laboratory, Ocean Optics and Remote Sensing Section | http://www7240.nrlssc.navy.mil/ocolor/Database/datab_experiment.html | CZCS, AVHRR and SeaWiFS products for many selected marine areas in the world |
| Naval Research Laboratory Stennis Space Center | http://www7300.nrlssc.navy.mil/altimetry/ | Several satellite images for many selected marine areas in the world |
| California Cooperative Oceanic Fisheries Investigation | http://www-mlrg.ucsd.edu/calcofi.html | Several surveyed oceanographic data for Eastern Pacific |
| GSFC/DAAC | http://xtreme.gsfc.nasa.gov/ | Worldwide atmospheric, hydrologic and oceanic datasets |

The introduction of the variety of oceanographic datasets in GIS is considered a separate task in GIS analysis and development. Vectorisation of aerial photography and hardcopy nautical charts, used in marine navigation, is a common practice in oceanographic GIS. In 1997, for example, collaboration between NOAA's Coastal Services Centre (CSC) and the Office of Ocean and Coastal Resource Management (OCRM) resulted in a regional ocean GIS, which serves as a unifying and non-contentious platform for regional ocean planning and policy dialogue in the southeastern US (Fowler and Gore 1997). The objective was to produce a high-resolution shoreline for the area by vectorising the scanned tide controlled photography used for NOAA nautical charting and integrating surveyed bathymetry up to the 200-mile limit of US exclusive economic zone (EEZ). The Oceanographic Analyst is a GIS extension, which was especially created for the Glacier Bay oceanography project, but is designed for use with any oceanographic dataset. The extension allows 3D and 4D analysis and display of volumetric and time series oceanographic data featuring several special modules for processing CTD data, calculating photic depth, integrating CHL data and processing weather data (Hooge *et al.* 2000). In 1996, the Marine and Fisheries Research Institute in Kenya (KMFRI) developed a GIS database comprising of existing information on biophysicochemical characteristics, resources, recreation and socio-economics (Ong'anda 1997). The data has been extracted from various project reports and theses written by KMFRI staff and external agencies. As a first stage in the GIS database management, metadata for all GIS layers have been catalogued in terms of the actual data holding, the time frame, units of measurement and feature type. Data representation assisted in effective dissemination of information through map representation facilitating visual impact and simplicity, especially among policy makers and managers. Tables accompany the data presented in map form. Some of the datasets have been packaged using various data formats in CDROM and diskettes.

Wilkinson (1996) and Hinton (1996) reviewed several issues in the integration of GIS and RS data for environmental applications noting the increasing need of using GIS in RS data storage, manipulation, and visualisation. Wright *et al.* (1997) thoroughly discussed the GIS interfacing of oceanographic data, particularly that of sidescan sonar imagery for deep-sea sampling and mapping. Also, Wright *et al.* (1998) developed a suite of conversion tools between a commercial GIS package and Generic Mapping Tools (GMT), a public domain software package for data manipulation and generation of high quality maps and scientific illustrations (Wessel and Smith 1995). Durand (1996) presented a GIS tool for the storage of various oceanographic data and Valavanis *et al.* (1998) introduced methodologies for the automated GIS interfacing of marine (oceanographic and fisheries) data. Davis-Lunde *et al.* (1999) presented a GIS-based processing and quality checking mechanism for data entering from NAVOCEANO's Data Warehouse (US Naval Oceanographic Office). The system handles both remotely sensed and *in situ* data. Livingstone *et al.* (1999) evaluated the involved cost of various oceanographic data sampling methods mentioning that there is always a trade-off involving the cost, spatial resolution and temporal control for each method and this trade-off ultimately determines the absolute accuracy and suitability of the imagery for the required purposes. Delarue *et al.* (1999) developed an Internet-based, client/server application to handle Autonomous Underwater Vehicle (AUV) data requests from users via a web browser. The web client displays AUV data through a GIS approach, merging the various data models in the web server. The AUV data organisation in GIS for storage, analysis and visualisation purposes was also used by Mallinson *et al.* (1997) in sea floor mapping surveys. Sowmya and Trinder (2000) reviewed some of the approaches used in research for image understanding of aerial and satellite imagery in GIS. Researchers from the fields of photogrammetry, RS, computer vision and artificial intelligence brought together their particular skills for automating tasks of information extraction from remotely sensed data. Dubuisson-Jolly and Gupta (2000) described a new algorithm for combining colour and texture information for the segmentation of colour images. Maximum likelihood classification combined with a certainty-based fusion criterion, the algorithm is used in updating of old digital maps (databases) of an area using aerial images.

Condal and Gold (1995) proposed a dynamic spatial data structure, based on Voronoi tesselation methods, as a potential method to specify the spatial relationships of unconnected objects in a marine GIS. Kitamoto and Takagi (1999) proposed two new probabilistic models, the Area Proportion Distribution (APD) and the Mixel Distribution (MD), for the classification of satellite images. They account the concept of mixels (mixed pixels with heterogeneous information), which originate from the fact that an image is the spatially quantised representation of the real world due to the finiteness of sensor resolution. Jingsong and Ying (1999) provided theoretic and technical support to the spatial and temporal analyses for modelling dynamic coastal areas. They focused on the solution of how to represent spatial attribute and temporal information in an integrated form using a data model on relational databases together with an object-oriented model for system development. Lienert *et al.* (1999) developed a software that collects and analyses backscatter data from a multiwavelength scanning Lidar system in real-time. The system can be used to process data from differential atmospheric absorption lidars and various oceanic lidars. De Lauro *et al.* (1999) described a GIS modular design for environmental data acquisition and for retrieving, preprocessing and visualising georeferenced information from oceanographic cruises. The system (OSIRIS: Ocean Survey Integrated Research Information System) has been tested and used in several seafloor mapping expeditions in the south-east Tyrrhenian margin (Italy) funded by the National Geological Survey of Italy. Knudsen (1999) developed a prototype of an integrated system for handling, analysis and visualisation of ocean data (Busstop). The system integrates functionality from GIS, image processing systems and database management systems for the synergistic use of satellite-based radar altimetry and ancillary data, enabling it to work as a tool for analysis of dynamic phenomena.

Water on the Web (WOW) is a production funded by the US National Science Foundation offering unique opportunities for high school and first year college students to learn basic science through hands on oceanographic activities, in the lab and in the field, by working with state of the art technologies accessible through a free website (http://wow.nrri.umn.edu/). Among other utilities, WOW offers real-time data for various lakes of St. Louis River in Minnesota and a set of web-based data visualisation tools, including the Profile Plotter (interactive line plots of CTD like data) and the DVT toolkit (interactive slices of several CTD like data). The whole service is based on an interactive GIS (NetWatch Science Magazine 2000).

Hatcher *et al.* (1997) and Hatcher and Maher (2000) discussed the use of GIS for real-time organisation of marine survey data. They discussed various methods of introducing vessel position, remotely operated vehicle (ROV), drifter and other oceanographic instrument data in real-time GIS applications. De Oliveira *et al.* (1997) presented a computational environment for modelling and designing environmental geographic applications focusing on users who are experts in their application domain, but who do not have adequate background in software engineering or database design. Bjorner (1999) presented a three-stage approach to software development for monitoring and decision support describing architectures for such systems, illustrating a fundamental approach to separation of concerns in software development and providing a proper way of relating GIS and demographic information systems to decision-support software. Paton *et al.* (2000) discussed the various spatial data models and software architectures that allow database systems to be efficiently used with spatial data. They presented an architecture for vector spatial databases that covers a range of typical GIS functions. Su *et al.* (2000) presented UCLA's (University of California, Los Angeles) GIS Database and Map Server, a cooperative effort of GIS interested parties on the UCLA campus, to deliver centrally stored geographic data via the World Wide Web (WWW). Terabytes of data are stored in a central GIS database into two ways, first using online hard disks for the most recently requested data units and second, using a large tape robot for less frequently requested data (http://gisdb.cluster.ucla.edu). Since the online disk space that is required for a large GIS server with terabytes of data is a significant cost, a tape robot can reduce the associated costs, if it is used as an automated digital tape library. Using a mechanical 'arm', the tape robot selects the requested tapes and places them in online drives. Marchisio *et al.* (2000) presented several optimised querying methods of RS and GIS repositories using spatial association rules. ANNEX I includes several macro routines for the introduction of oceanographic data into a GIS environment.

## 2.7 IDENTIFICATION AND MEASUREMENT OF UPWELLING EVENTS

Marine ecosystems are supplied by nutrients that are recycled within the euphotic 'surface' zone by influx from intermediate and deep waters. When the wind blows over the surface of the ocean it sets up a stress, which causes the water to move in the same direction as the wind. Once the water starts moving, it is affected by the Coriolis force that is a result of the earth's rotation (earth's spin). This causes direction of the water movement to be deflected by 90° either to the right (northern hemisphere) or left (southern hemisphere) of the wind direction (Ekman transport). If the Ekman transport is set up away from the coast, the surface waters are moved offshore and are replaced by deeper cold water that is upwelled close to the coast. This process is called wind driven upwelling with bottom and coastal topographies playing important roles (Figure 2.1). The upwelled water is usually from below the pycnocline (deep water masses) and so is nutrient rich. Microscopic plants of phytoplankton, which float along the surface, use the nutrients of the upwelled

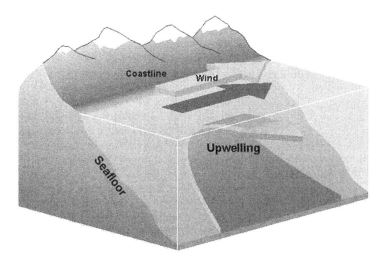

**Figure 2.1.** Typical wind driven coastal upwelling. Wind blowing parallel to the coast forces surface water to move away from the coastline and cold nutrient rich deep water to upwell.

water as food and bloom. Because of the increase in productivity of upwelling areas, which take about 0.1 per cent of the ocean surface area, it is found that they account around 50 per cent of the world's fisheries.

There are seven major upwelling areas in the world oceans (Figure 2.2). Each of these areas is an area of very high biological productivity. They are major upwelling areas due to the predominant equatorward winds, which blow parallel to the coast generating broad shallow eastern boundary currents that combine to be extremely favourable to the upwelling process. Upwelling events are not only found in these major areas. Many other places in the world oceans are characterised by localised smaller scale upwelling, which occur on most continental shelves, characterising local scale productivity dynamics and often determining local fishery activities.

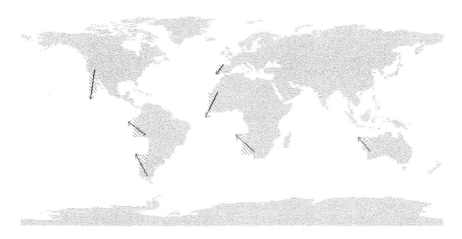

**Figure 2.2.** Major upwelling areas in world oceans with arrows showing dominant wind direction.

Upwelling is monitored in many world regions. NOAA's Pacific Fisheries Environmental Laboratory (PFEL) generates indices of the intensity of large scale, wind induced coastal upwelling at 15 locations along the California Current System, 11 locations along the Peru/Chile Current region, 13 locations along the Canary Current, and 7 locations along the Benguela Current. The indices are based on estimates of offshore Ekman transport driven by geostrophic wind stress. Geostrophic winds are derived from six hourly synoptic and monthly mean surface atmospheric pressure fields provided by the US Navy Fleet Numerical Meteorological and Oceanographic Centre. The idea behind the upwelling indices is to develop simple time series that represent variations in coastal upwelling. Daily and monthly index time series are provided regularly to scientists and managers concerned with marine ecosystems and their biota. Such indices have been used in many studies and scientific publications. PFEL also provides an online 'Live Access Server' with global upwelling indices (http://www.pfeg.noaa.gov/). The Coastal Ocean Observation Lab at Rutgers University studies upwelling events offshore New Jersey using CDT, measured wind data and satellite images of SST. Upwelling in New Jersey coastal area is characterised by convergence zones. When winds continue blowing from the southwest after an upwelling has begun, the upwelling areas tend to converge around consistent locations along the various inlets of New Jersey's coastline. It is believed that the coastal upwelling band begins to converge into small, weak eddies at approximately 50 km intervals. Temperatures at the beach can then vary up to 10 °C.

Solanki *et al.* (1998) used AVHRR SST imagery to study various features of the upwelling process in the Arabian Sea. The event is observed along the Arabian coast during pre-southwest monsoon (June–July) and along the Gujarat coast during post-southwest monsoon (September-October) with benefits to local fishing operations. Li and Shao (1998) proposed a theoretical GIS estimation model and its resolution method for the spatial calculation of oceanic primary productivity. The method removes the atmospheric aerosol effects on CHL concentrations derived from remotely sensed ocean colour data. Su and Sheng (1999) integrated a GIS and multidimensional visualisation software to visualise CTD and bathymetry data in the Monterey Bay, California for a better understanding of the upwelling processes. The GIS system performs data interpolation, unifies map projection and filters the processed data to a computer visualisation package, in which the centre and the maximum depth of upwelling are identified through a series of animations. Valavanis *et al.* (1999) developed UPWELL (Upwelling Identification and Measurement System), which is a set of GIS routines that use vector analysis on satellite images of SST and CHL concentration to identify and measure several characteristics of upwelling events. Demarcq and Faure (2000) used thermal infrared data from the satellites of the European Meteosat series to characterise the dynamics of the West African coastal upwelling. Images were used to characterise the spatial structure of the upwelling by automatic localisation of the SST minima at selected coast locations and to derive a normalised upwelling intensity index based on SST differences, which were quantified in terms of intensity and seasonal lag.

GIS contribute to the study of the upwelling process by integrating satellite images and by providing a visual time series of several properties (e.g. SST and CHL concentration) for a region. This integration allows users to see the ocean surface feature evolve both in time and space. In a simple 2D GIS setting, time series of georeferenced satellite images of SST and CHL concentration, satellite or measured wind and sea current data and bathymetry may be integrated and provide GIS output on certain measurements of the upwelling process. Depending on the temporal extent of these data, GIS analysis reveals the epicentre of upwelling events, their duration, size and strength (by means of the resulted temperature and CHL variation). Also, in a time series of events, GIS calculate the average wind and sea current patterns that trigger the upwelling process. A proposed GIS methodology includes the following data processing: Each SST grid that contains cold patches of water is automatically classified by vector polygons based on a 1 °C classification table. Polygons that describe the cold patches of water as well as the surrounding area are separately selected and used for the isolation of the associated grid values, which are saved in new grids. The difference in SST variation and the cold patch epicentre (lowest cell value: lowest temperature) are calculated from the new grids. The patch area is calculated from the selected vector polygons. Then, validation by wind data is performed. Wind measurements are averaged in weekly or monthly force means for all four-wind directions (N, S, E, and W). Integration between epicentre location and wind data is performed by overlaying wind measurements on each upwelling location, thus checking for compatibility between location and wind direction based on upwelling theory. Then, wind verified cold SST patches are characterised as upwelling. Finally, as in SST grids, the vector polygon technique is applied to CHL images for the measurement of CHL variation. The GIS methodology for the study of the upwelling process is applied to two examples. In the first example, two average satellite images of SST (AVHRR) and sea surface CHL concentration (SeaWiFS) of February 2000 may be integrated in GIS and reveal the effect of the upwelling process to temperature and CHL variation off SW African coast (Figure 2.3). The second example examines a coastal area of small upwelling events in North Aegean Sea in SE Mediterranean by integration of weekly AVHRR SST and measured wind data for the period from May 1993 to December 1995 (Figure 2.4). The second case reveals the epicentre of these small upwelling events, their duration, size, and strength in temperature variation. Proposed GIS routines for such measurements are supplied in Annex I.

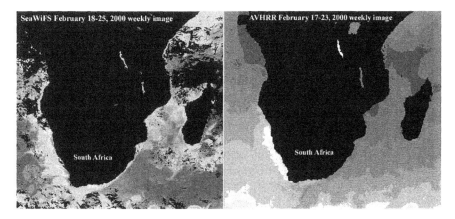

**Figure 2.3.** Seasonal large upwelling off SW African coast. The extent of the upwelling event is shown by increase in sea surface CHL concentration (from SeaWiFS) and decrease in SST distribution (from AVHRR). Further processing of such images in a GIS environment results in certain measurements on the upwelling process.

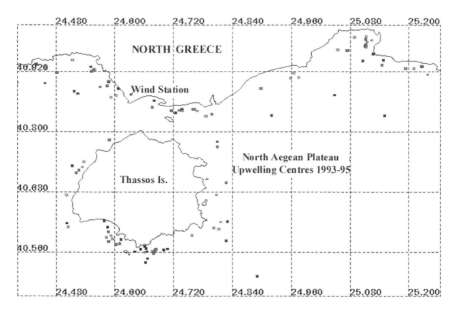

**Figure 2.4.** Upwelling centres at the North Aegean Plateau (Eastern Mediterranean) derived from weekly AVHRR SST images for the period 1993–1995.

## 2.8 IDENTIFICATION OF TEMPERATURE AND CHLOROPHYLL FRONTS

Oceanic fronts are areas of particular interest contributing to the continual nutrient mixing in the oceans. In oceanic front areas, there is rapid change in temperature, CHL and salinity distribution while the horizontal gradients of these properties are homogeneous in the surrounding water masses (Figure 2.5). Usually, these events take the form of long stripes on the sea surface and may be connected to other oceanographic processes, such as upwelling and tide currents. Fronts have significant effects on biology. These systems tend to form zones of convergence of different water masses resulting in accumulation of planktonic organisms. This aggregation also affects the distribution of secondary producers and pelagic predators.

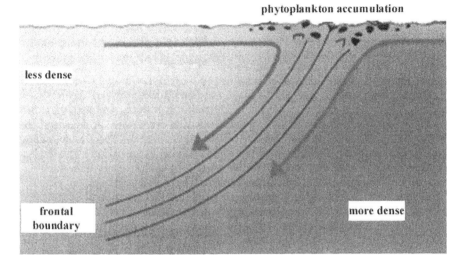

**Figure 2.5.** Vertical profile of a typical front boundary between more and less saline water. Productivity increases at fronts due to accumulation of phytoplankton.

Waters of the continental shelf tend to be slightly less saline (due to run-off and river water input) than the rest of the ocean being also more prone to heating and cooling. In the boundary between continental shelf and open ocean waters exist fronts marking the transition from shelf waters to the open ocean (shelf break front). In winter, cooling and wind mixing mean that the shelf water is cooler and less saline than the warm saline water offshore. The front becomes an area of high production as developing sea currents enhance nutrients in the surface waters. The results are increase in production and fish aggregation. Ocean frontal boundaries are located throughout the world's oceans. They represent areas of high interest to scientists of many disciplines and to commercial fisheries industry because they represent regions of strong anomalies in the ocean, high biological activity, dynamic chemical processes and change in acoustic propagation. Some oceanic fronts are global in extent, such as the Subtropical Front, which extends around the Southern Ocean marking the boundary between the Subantarctic Surface Water and the Subtropical Surface Water masses.

The use of satellite data in ocean front studies is common. In United Kingdom, Earth Observation Sciences (EOS) maps and analyses the boundaries between ocean water masses through the combined manipulation of AVHRR,

ATSR and SAR data providing products to many academic institutions and the fishing industry. The Colorado Centre for Astrodynamics Research, in cooperation with NASA, uses satellite altimeter derived data to detect ocean fronts. Altimetry data is combined with bathymetry, water temperature and historical fish catch statistics in GIS, producing maps of high interest to commercial fishing industry. NAVOCEANO (the US Naval Oceanographic Office) generates products relating to oceanographic fronts, which are provided to the naval fleet and to commercial enterprises via hardcopy and online maps. The programme 'Ocean Fronts: Their contribution to New Zealand's marine productivity' is a 4-year attempt (started in 1999) on gaining an improved understanding of what drives the marine ecosystem in the region of Southern New Zealand where the Subtropical Front is dominant. This global front becomes narrow in this region following the region's bathymetry patterns mixing the warm, saline, nutrient poor subtropical waters with the cool, less saline, nutrient rich subantarctic waters. The project focuses on physical variability and phytoplankton dynamics of the front (O'Driscoll and McClatchie 1998). Askari *et al.* (1996) developed a system that uses remotely sensed data for the detection and classification of oceanic fronts in subsequent images. Mesick *et al.* (1998) used AVHRR SST images and ESRI ARC/INFO GRID GIS for the automated detection and mapping of oceanic front and eddies in the area of North Atlantic where formations of Gulf Stream's rings and eddies are dominant. They also integrated bathymetry to create a projected slope grid, a gradient temperature grid derived from the slope and bathymetry grids and finally a line coverage from gradient grid representing the front. Menon (1998) also used SST satellite images to describe the important role that oceanic fronts play in productivity variations in the Southern Indian Ocean. Moore *et al.* (1999) used AVHRR SST data for the study of the Antarctic Polar Front (APF). Gradient maps (images showing regions of strong horizontal gradients in SST, every 1.35 °C over 45 to 65 km) were constructed from the SST images. The combination of the SST images and the gradient maps was used as the basis for the digitisation of the pole-ward edge of APF. Waluda *et al.* (2001) also used SST images in GIS and derived gradient values for the identification of fronts in SW Atlantic. Areas of high SST gradient were identified by calculating the range of values within a series of 3x3 pixel matrices and applying a threshold range of 0.4 °C to 1 °C for the SST thermal gradients (fronts). Shaw and Vennell (2001) studied the spatial and temporal variability of the global Subtropical Front using AVHRR SST images. They developed an algorithm that calculates the position of the front, the mean position of the front over a period of time and various SST front parameters (mean SST, SST difference and SST gradient). Results were compared with CTD data proving the algorithm's results for front mapping.

Image pattern recognition methods to measure ocean surface movement using sequential satellite images have been widely developed (May 1993; Cummings 1994; Passi and Harsh 1994) while several of these methods are specifically applied to oceanographic fronts (Khedouri *et al.* 1976; Suzanne and Lybanon 1983; Lybanon 1996). Wai *et al.* (1994) developed a statistical edge detection algorithm for the selection of ocean thermal patterns by detecting and mapping gradients from satellite images. The algorithm locates the best match to the pattern in a subsequent image and then, surface displacement direction and distance are calculated for each feature.

Extraction of anomalies in SST distribution (or CHL concentration) is another method for mapping oceanic fronts. The spatial and temporal extents of the boundaries of such anomalies are good indicators of persistent front systems. Extraction of SST anomalies using time series of monthly (or weekly) SST images (e.g. for a 5- or 10-year period) may be calculated by averaging SST values for each month (or week) of the year and then subtracting an average month (or week) from the corresponding month (or week) in the time series (e.g. Jan98_anomaly = Jan98 – Jan_average). The regions showing persistent anomalies indicate locations of oceanic fronts.

Two examples from SW Atlantic Ocean (Figure 2.6) and SE Mediterranean (Figure 2.7) show the distribution of fronts in temperature distribution. In both cases, AVHRR SST imagery was georeferenced and analysed in GIS. In the case of SW Atlantic, a simple edge detection algorithm was applied to the images, while in SE Mediterranean SST anomalies were calculated for the period March 1993 to December 1999.

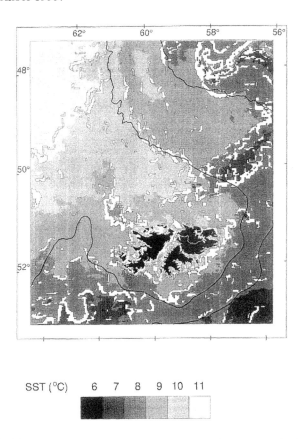

**Figure 2.6.** GIS derived fronts from AVHRR satellite images of SST in SW Atlantic (Falkland Islands/Islas Malvinas). Image is courtesy of Claire Waluda, British Antarctic Survey, UK (Waluda *et al*. 2001).

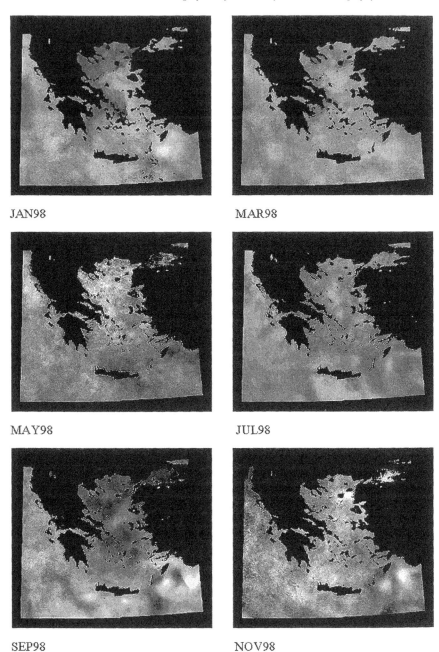

**Figure 2.7.** Sea surface temperature (SST) anomalies in SE Mediterranean derived from AVHRR
monthly imagery for the period 1993–1999. Lighter areas show stronger SST anomalies and indicate
the existence of persistent fronts.

## 2.9 IDENTIFICATION AND MEASUREMENT OF GYRES

Basin circulation consists of major ocean gyres that circulate clockwise (anticyclonic gyre) in the northern hemisphere oceans and anticlockwise (cyclonic gyre) in the southern hemisphere oceans. These processes are known as subtropical gyres. These gyres are driven by global wind patterns that are created by uneven heating of the earth's surface by the sun and by the Coriolis force (earth's rotation). Besides large ocean gyres, smaller basins like the Mediterranean Sea, are characterised by seasonal gyre formations. Some of these gyres may be viewed as another type of upwelling. Often, these processes are open sea upwelling events of dynamic pumping of bottom water to the surface. These events are stimulated by complex wind and sea current patterns with bottom topography playing an important role. Usually, gyres are characterised by certain seasonality and vary in duration, strength and size. At the boundaries of the gyres exist strong geostrophic currents. Depending on the direction of the process, a gyre may create a region with warmer (cyclonic gyre) or colder (anticyclonic gyre) SST. These processes can be identified through AVHRR SST images because of their difference in temperature as compared to the surrounding area. Figure 2.8 shows the major gyre formations in the Mediterranean Sea.

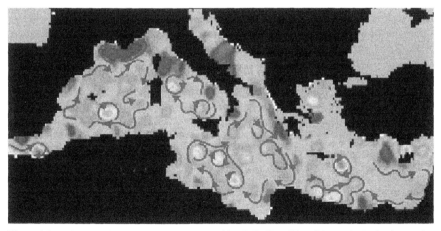

**Figure 2.8.** Major gyres in the Mediterranean Sea with TOPEX/Poseidon altimetry in the background.

Monitoring of gyre activity in the oceans becomes important because it reveals surface geostrophic currents, seasonal CHL concentrations and seasonal offshore fish feeding grounds. For example, the Department of Ocean Development (Government of India) conducts experimental and operational studies on seasonal and annual formation and propagation of small-scale gyres using thermal satellite data. Thorpe (1998) overviewed the 40-year knowledge about our oceans, which derived mainly from observations of turbulence in the stratified and rotating world ocean since the 1960s, when mesoscale motions with scales of 30–150 km and 100 days were discovered by neutrally buoyant floats, to the 1990s and the use of SF6 'purposeful tracer' release. Most of the ocean is stably stratified but it contains a rotational turbulent continua and isolated rotating eddies and gyres and Rossby waves. The presence of lateral boundaries of the continental land masses, islands and seamounts provides constraints to the circulation and to the propagation of eddies and gyres and possibly substantial sources and sinks of eddy motion. Also, channels connecting the oceans to land locked seas (e.g. the Mediterranean) result to formation of water with anomalous properties that act as natural tracers (e.g. temperature or salinity allows interthermocline eddies to be readily detected). In addition, convection and differential seasonal latitudinal forcing introduce upper ocean variability and intrathermocline eddies. Sharma *et al.* (1999) identified marine regions of high energies associated with various current systems under the influence of monsoon winds in the Arabian Sea and the Bay of Bengal. They used TOPEX altimeter data to calculate the monthly mesoscale eddy kinetic energy per unit mass in the region, showing that the mesoscale eddy kinetic energy is highest near the Somali region during the SW monsoon due to formation of mesoscale eddies and also because of upwelling. In the Bay of Bengal, high eddy kinetic energy is seen toward the western side during non-monsoon months due to the western boundary current. Karl (1999) used a decadal time series of oceanographic data to study the physical and biogeochemical processes in the North Pacific Subtropical Gyre, the largest ecosystem on our planet. Burrows and Thorpe (1999) used observations acquired by 42 ARGOS-tracked drifters to study the mean flow circulation patterns around the Hebrides and Shetland Shelf slope (NE Atlantic).

Dijkstra and Molemaker (1999) studied the wind driven gyres in the North Atlantic Ocean by determining the structure of steady solutions within a hierarchy of equivalent barotropic ocean models. The findings demonstrated that symmetry breaking is at the origin of two different mean states of the Gulf Stream related to a small number of oscillatory modes of either interannual or intermonthly periods. Napolitano *et al.* (2000) used a coupled physical and biological model to study the seasonality of biological production characteristics of the gyres at the Rhodes and western Ionian basins (Eastern Mediterranean). Drakopoulos *et al.* (2000) used a GIS for the mapping of a seasonal gyre in SE Mediterranean. Valavanis *et al.* (2000) developed a GIS-based application to derive seasonal measurements for the West Cretan Gyre. Gomis *et al.* (2001) used a multivariate optimal statistical interpolation method, which was applied to conductivity/temperature/depth (CTD) and ship mounted acoustic doppler current profiler (ADCP) data aiming to improve the spatial interpolation of any particular variable (e.g. dynamic height) by including in the analysis observations of other physically related variables (e.g. current). Harrison *et al.* (2001) used various data from oceanographic surveys to describe the large-scale variability in hydrographic, chemical and biological properties of the upper water column of the subtropical gyre and adjacent waters in North Atlantic Ocean, contributing to recent efforts to partition the ocean into distinct biogeochemical provinces.

The following example illustrates the mapping of the epicentre movement of the West Cretan Gyre (WCG) in SE Mediterranean (Figure 2.9). The GIS output also includes establishment of the gyre's seasonality and strength by means of temperature variation (see also Colour plate 2). For this purpose, a time series of weekly AVHRR SST imagery were georeferenced in a GIS database for the period from March 1993 to December 1997. As in the case of upwelling, vector polygons are used to define the area of the gyre using the distribution of SST inside and outside the gyre process. Each AVHRR grid is automatically classified by vector polygons based on a 1 °C temperature classification table. Then, polygons that describe the area of the gyre as well as its surrounding area are selected and used for the measurement of the gyre formation. The activity of WCG is associated with colder surface temperatures and higher CHL concentrations than in normal conditions because the cyclone functions like a pump of bottom colder and nutrient rich water to the surface.

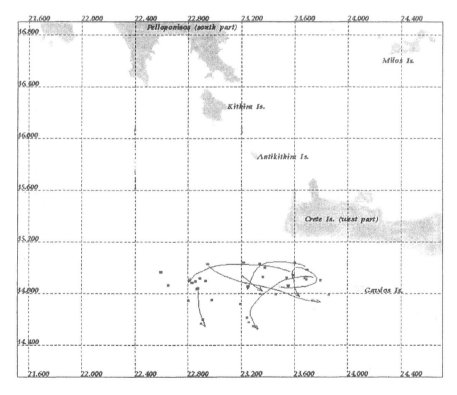

**Figure 2.9.** GIS mapping of a seasonal gyre in SE Mediterranean: Movement tracks of the West Cretan Gyre epicentres (area of coldest SST in the gyre) during the period March 1993 to December 1997.

## 2.10 CLASSIFICATION OF SURFACE WATERS

The classification of surface waters is an important parameter for the study of the upper layer of the euphotic zone. The classification approach provides information on certain classes of sea surface waters, which are characterised by certain value ranges in specified environmental parameters, such as temperature, CHL and salinity. The spatial extent and temporal change of these value clusters provide important information on the seasonal mixing and productivity levels of sea surface waters. Persistent oceanographic processes and the effects of these processes to the distribution of certain environmental parameters may be revealed. In connection to fisheries production data, these clusters of surface water are an indication of fishing grounds and species habitats. Nowadays, RS databases go back about two decades and are expanding rapidly while the quality of the data is continually increasing. Scientists can now conduct long-term analyses to extract seasonal, annual or climatological information.

Through various hydrological studies on bays and estuaries, the Texas Parks and Wildlife Service Department of the US Coastal Studies Programme, Resource Protection Division, identified seasonal salinity zones at Galveston Bay and the Trinity/San Jacinto Estuary in the Gulf of Mexico. GIS techniques were used to compare salinity maps generated by a hydrodynamic model under optimised maximum and minimum fresh water inflows. These maps were generated by contouring the salinities from each model run using GIS-based interpolation methods. Salinity change analysis was performed by overlaying minimum and maximum monthly salinity maps, thus producing salinity difference maps. In addition, these maps were used as basic data coverages on overlays with fisheries catch data for the identification of preferred salinity zones for seven target species (white and brown shrimp, blue crab, Gulf menhaden, Atlantic croaker, bay anchovy, and pinfish). From these data integrations, two critical data values for each species were calculated (the percent abundance of animals in bay salinity zones and the percent of bay area occupied by that salinity zone). Maps from this water classification work are available at: http://www.tpwd.state.tx.us/texaswater/sb1/enviro/galvestonbay-trinitysanjac/inlandflow.html.

Christensen *et al.* (1997) developed an index of biological sensitivity to changes in freshwater inflow for 44 species of fishes and macroinvertebrates in 22 Gulf of Mexico estuaries. The BioSalinity Index (BSI) provides an innovative approach to quantify the sensitivity of organisms to changes in estuarine salinity regimes based upon known species salinity habitat preferences, the availability of this preferred habitat and the relative abundance and distribution of species in time and space. Gao and O'Leary (1997) used colour aerial photographs and *in situ* water samples to classify the waters in the Waitemata Harbour (Auckland, NZ) according to the total mass of suspended solids in the area. Kitsiou and Karydis (2000) studied the spatial distribution of eutrophication in the marine environment using the inverse distance weighted (IDW) interpolation method on phytoplankton community data in a GIS. They mapped four different trophic levels in the marine environment of Saronicos Gulf in Greece (eutrophic, upper mesotrophic, lower mesotrophic, and oligotrophic). Sklar and Browder (1998) overlaid salinity patterns onto habitat features for calculating area of overlap and measuring spatial patterns of salinity change in response to variation in freshwater inflow, thus determining the spatial extent of specific salinity/habitat combinations under various scenarios of freshwater inflow in the Gulf of Mexico. Drakopoulos *et al.* (2000) integrated a 2-year time series of SeaWiFS and AVHRR images and salinity climatology for the Greek Seas (SE Mediterranean) to seasonally classify surface waters in the area. They produced four images that show the seasonal change in geographic distribution of similar surface water masses.

The following examples include the classification of Greek surface waters in SE Mediterranean (Figure 2.10) and the classification of NW Atlantic surface waters (Figure 2.11). North-west Atlantic waters were classified by Huettmann and Diamond (2001) in an extensive GIS development for determining seabird distribution based on environmental variables. South-east Mediterranean surface waters were classified for their temperature, CHL and salinity seasonal contents. For this purpose, a time series of monthly AVHRR and SeaWiFS imagery (from September 1997 to August 1999) and MODB (Mediterranean Oceanic Database) salinity climatologic dataset (1980–1989) were georeferenced and analysed in a GIS environment. First, monthly average grids for each environmental parameter were calculated from the data time series. Second, seasonal averages were calculated from the monthly averages (winter, spring, summer, fall). The resulted 12 grids were placed in four grid stacks, each containing seasonal grids for each environmental parameter (temperature, CHL and salinity). Unsupervised classification of these grid stacks revealed four distinct clusters that describe the dynamics of sea surface waters in Greek Seas on a seasonal basis.

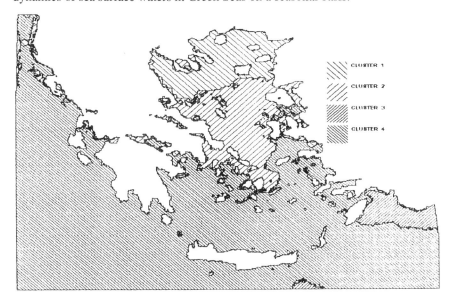

**Figure 2.10.** Classification of SE Mediterranean surface waters according to September 1997 to August 1999 average CHL concentration and temperature/salinity distribution. Cluster values are as followed:
Cluster 1:0.19 mgr$^{-3}$, 18.4 °C, 36.3 psu, Cluster 2:0.10 mgr$^{-3}$, 19.0 °C, 38.4 psu
Cluster 3:0.08 mgr$^{-3}$, 20.3 °C, 38.8 psu, Cluster 4:0.06 mgr$^{-3}$, 22.0 °C, 38.8 psu

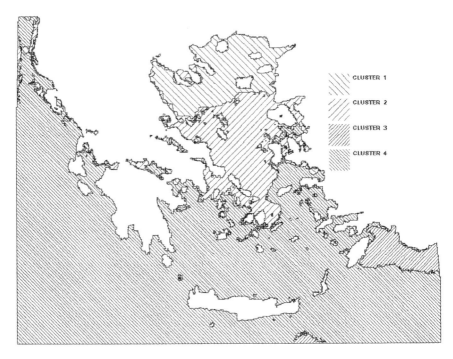

**Figure 2.11.** GIS classification of surface waters and spatial distribution of 5 clusters of 20 environmental data layers in various seasons in NW Atlantic Ocean. The major data layers that were used for this classification include air temperature and pressure, wind speed, sea depth, slope and aspect of sea floor, SST and salinity and sea temperature and salinity at 30 m of depth (Huettmann and Diamond 2001). Images are courtesy of Falk Huettmann, Simon Fraser University, Canada (http://www.sfu.ca/biology/homepage.html).

## 2.11 MAPPING OF THE SEABED

The combined mapping of coastal bathymetry and benthic habitats is widely used as baseline information for the understanding and management of coastal environments. Various substrate types support different benthic habitats while bathymetry often plays the role of vegetation zoning factor. Information on the structure of seabed is a very important parameter for the study of coastal processes as well as for the identification of marine species spawning grounds. For this purpose, the application of marine geophysics and GIS techniques to the characterisation of benthic habitats has increased our overall knowledge about our sensitive coastal zones and the ability of fisheries managers to assess species distribution and habitat types. Agenda 21, the comprehensive action plan adopted by the UN Conference on Environment and Development (UNCED) in 1992, encourages the identification and conservation of marine ecosystems exhibiting high levels of biodiversity and productivity. The various types of benthic habitats (e.g. *Macrocystis*, *Zostera*, *Posidonia*, etc.) are known worldwide for their diversity and production. They provide shelter for many benthic species and act as nursery grounds for many fish and other marine species. The spatial distribution of such habitats is a strong indicator of coastal ecosystem response to environmental change.

Coastal mapping is commonly performed with the combined use of on-site and remote methods. The common RS options for such mapping projects are aerial photography, SPOT (System Pour l'Observation de la Terre) and Landsat Thematic Mapper (TM) imagery. The use of aerial photography is relatively less expensive and enables the choice for appropriate weather and sea conditions, tide and sun angle to ensure high quality images. In addition, the hyper spectral imaging (e.g. AVIRIS and CASI), the laser fluorosensing and LIDAR (light detection and ranging) are three airborne sensing methods widely used in coastal benthic mapping. Crossney (2000) discussed methods of LIDAR data accuracy control and underlined the need of such data in real-time on beach conditions for coastal management. Also, various sonar-based methods are widely used (e.g. sidescan, multibeam). Systems of acoustic signal processing (e.g. RoxAnn®) are repetitively used in submerged aquatic vegetation mapping due to low cost, large area coverage and easiness to use. However, such systems require extensive calibration and ground truthing.

Although coastal habitat mapping is generally expensive (depending on the extent of the sampled area), RS is a more cost-effective technique than alternative field survey methods. Mumby *et al.* (1999) investigated the various required costs for using remotely sensed data in coastal habitat mapping. They identified four types of costs when undertaking RS (set up costs, field survey costs, image acquisition costs and the time spent on analysis of field data and processing imagery). Their analysis showed that SPOT imagery is the most cost-effective method for mapping small coastal areas (less than 60 km in any direction) in coarse detail while Landsat TM is the most cost-effective and accurate sensor for larger areas. Generally, very detailed habitat mapping should be undertaken using digital airborne scanners or interpretation of colour aerial photography.

Basu and Saxena (1999) reviewed the advantages and disadvantages of various data collection methods used by nine shallow water sonar mapping systems of coastal, nearshore and other shallow water regions, which have been tested in the real environment and are in use around the world. They describe the four main categories of sonar systems (single beam, multibeam forming, multibeam interferometric and side scan sonars) concluding that the systems, which combine side scan sonar with bathymetry are more useful in the identification of bottom features and analysis of seabed morphology. Clarke (1999) compared five swath sonar systems in terms of the output data formats and the needs in data transformations for further analysis. Gorman *et al.* (1998) overviewed field methods, data collection and analysis procedures applied to four key coastal datasets used in monitoring and baseline studies: (1) aerial photography; (2) satellite imagery; (3) profile surveys; and (4) bathymetric (hydrographic) records. NOAA's Coastal Services Centre (CSC) developed a freely available CDROM entitled 'Submerged Aquatic Vegetation: Data Development and Applied Uses', which contains an extensive number of examples and methodological procedures on how data on submerged aquatic vegetation (e.g. aerial photography) are being used for benthic habitat GIS mapping (http://www.csc.noaa.gov/crs/bhm/).

The most common and accurate method for coastal sediment mapping is the combination of data from hydroacoustic surveys, underwater videography and aerial photography. In coastal areas where light penetration is high, the use of only aerial surveys is often adequate. For example, the benthic habitats of the Florida Keys were mapped using mainly aerial photographs. In 1998, collaboration among the Florida Department of Environmental Protection, NOAA National Ocean Service (NOS) and ESRI produced the Benthic Habitats of the Florida Keys CDROM, which is freely available by NOS Internet site (http://spo.nos.noaa.gov/). The project relied primarily on high-resolution aerial photography for the development of a GIS database. Underwater videography and singlebeam acoustics (e.g. RoxAnn® system) were used in turbid or deep-water areas beyond the euphotic zone. Sonar, benthic photography and GIS were also used for the mapping of benthic habitats off the Virginia coast (Cutter *et al.* 1998). In Greece, a 2-year mapping effort (1999–2000) among several marine centres and institutes resulted in the coastal habitat mapping of 67 NATURA sites located in the Aegean and Ionian Seas (SE Mediterranean). The data that were used in the mapping were sonar RoxAnn® data and aerial photographs, which were integrated in a GIS (Siakavara *et al.* 2000). The Grant F. Walton Centre for RS and Spatial Analysis at Rutgers University has developed a series of maps showing submerged aquatic vegetation in the Barnegat Bay region from 1968 to 1999 (Figure 2.12) applying geospatial technology (GIS and RS) for landscape and watershed ecological analysis (Lathrop *et al.* 2000). The Division of Marine Research of the Commonwealth Scientific and Industrial Research Organisation (CSIRO) in Australia used satellite imagery and aerial photographs, which were integrated in a GIS for the mapping of marine habitats in SW Australia. In 1995, the threatened population of manatees in Puerto Rico was the reason to identify habitats and develop criteria and biological information important to manatees' recovery. Computer-aided mapping based on the interpretation of aerial photographs and field ground truthing was used to define these habitats and map their distribution in the area of high manatee use (Kruer 1995). Through NOAA's Biogeography Programme, bottom sediment mapping in the Caribbean and Hawaii included many interrelated activities (e.g. multispectral image analysis applied to aerial photos, visual interpretation of scanned and hard copy photos, comparison of various classification techniques) resulting in the creation of a GIS bottom sediment database.

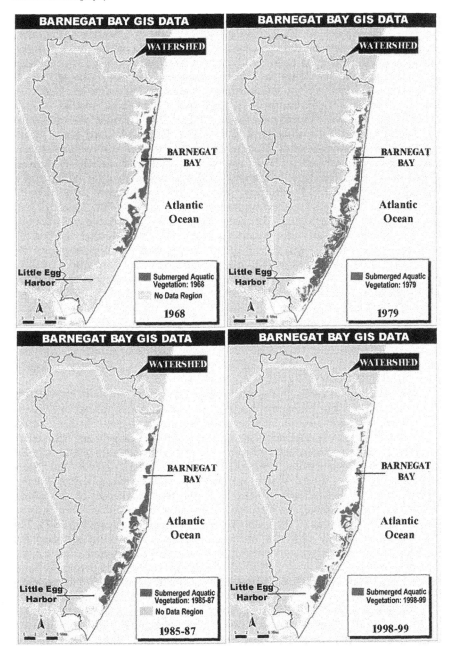

**Figure 2.12.** GIS mapping of submerged aquatic vegetation at Barnegat Bay, New Jersey using RS data. Differences in vegetation spatial distribution are shown for the period from 1968 to 1999. Images are courtesy of Rick Lathrop, Rutgers University, Center for Remote Sensing and Spatial Analysis (CRSSA), New Jersey, USA (see also at: http://www.crssa.rutgers.edu/projects/runj/bbay.html).

Caloz and Collet (1997) reviewed the methodological aspects of using GIS and RS in aquatic botany. They mentioned that the use of these new technologies in submerged vegetation mapping integrates a variety of disciplines, such as digital cartography, modelling of geographic space (e.g. geostatistics) and digital image processing and analysis. There are a diverse number of methodological approaches in bottom substrate mapping, which usually depends on the nature and quality of the surveyed data. Lehmann *et al.* (1994) used Generalised Additive Models (GAM) and GIS to model and map the distribution of submerged macrophytes in the littoral zone of Lake Geneva in Switzerland. They underlined that GIS and GAM appear as powerful tools to proceed from the description of species response curves to environmental gradients toward the spatial predictions of species distribution under changing environmental conditions. Donoghue *et al.* (1994) incorporated Landsat TM imagery into a GIS to compare two different classification techniques for mapping the intertidal zone of the Wash estuary. They used a conventional maximum likelihood classifier and a fully constrained linear mixture model with the later appearing to be the most accurate method for mapping pioneer zone vegetation. Bulthuis (1995) used aerial photographs and extensive ground truthing surveys to map the distribution of seagrasses in Padilla Bay in Washington. Data from ground surveys were used for the interpretation and classification of the aerial photography. Ferguson and Korfmacher (1997) integrated aerial photography and satellite imagery in a GIS to map seagrass meadows in North Carolina. They also integrated bathymetry data and found that seagrasses were present at depths below 2 m and especially abundant at depths of less than half a metre. Davies *et al.* (1997) developed GIS methods of analysing data from acoustic ground discrimination systems in conjunction with biological information to produce biological resource maps.

Bushing (1997) used GIS to perform gap analysis in order to identify gaps in coverage between the geographic distribution of species and habitat for defining the spatial configuration of marine reserves around Santa Catalina Island, California. Field survey data, aerial photography, satellite imagery and bathymetry were integrated using advanced GIS techniques to derive measures of bottom relief (depth contours, aspect, slope, substrate heterogeneity, habitat diversity and habitat fragmentation) from digital terrain models. The analyses indicated that existing reserves do not protect kelp resources on the windward side of the island. Schmieder (1997) mapped the submersed vegetation of the whole littoral zone of Lake Constance by boat on the basis of aerial photographs acquired in 1993. The maps were compared with two previous surveys (1967, 1978) showing that changes in aquatic macrophytes were highly related to the changes in trophic status of the lake. Pasqualini *et al.* (1998) identified that the wide bathymetric range (0–50 m) of *Posidonia oceanica* increases the difficulties involved in its mapping, noting that the combined use of image processing of aerial photographs for shallower layers (0–15 or 20 m in regions of shallow and sheltered waters) and of side scan sonar for deeper regions (20–50 m) is a particularly suitable approach. Hill *et al.* (1998) evaluated the propagation of the toxic green alga *Caulerpa taxifolia* in northwest Mediterranean Sea by means of a GIS-based algorithmic computer model. The model is capable of spatially predicting local patterns of alga expansion, increase of biomass and covered surfaces and invasive behaviour towards existing communities. Garrabou (1998) used GIS to create a composite image by merging georeferenced specimen outlines of benthic encrusting clonal organisms from different time steps. The method provided automatic and reliable data for estimating independent growth (net area gained) and shrinkage rates (net area lost) of benthic clonal organisms for any time period. McRea *et al.* (1999) used a GIS with high-resolution sidescan sonar data to determine various marine habitats in the area offshore of Kruzof Island in Alaska. Bell *et al.* (1999) used a raster GIS to investigate gap dynamics within a shallow subtidal landscape characterised by seagrass vegetation and to examine the relation between gap formation and selected physical factors in Tampa Bay, Florida. They used time series of presence/absence data of seagrass and compared maps between dates revealing the location and size of gap birth (origin) and extinction (closure). Haltuch and Berkman (1999) integrated sidescan sonar data (substrate type and bathymetry), distance from shore and change through time in years into a GIS that was used to model percent cover of exotic mussels (*Dreissena* sp.) on Lake Erie. The GIS model presented strong evidence of expanding *Dreissena* populations in the lake during 1994–1998 demonstrating that watersheds, which are invaded by these exotic mussels, evolve in extremely dynamic benthic habitats. Guan *et al.* (1999) evaluated different spatial interpolation methods for mapping submerged aquatic vegetation in the Caloosahatchee Estuary, Florida. They used bottom depth and seagrass height and density point data and applied three interpolation methods: (1) ordinary kriging with five different semivariance models combined with a variable number of neighboring points; (2) the inverse distance weighted (IDW) method with different parameters; and (3) the triangulated irregular network (TIN) method with linear and quintic options. The study identified that kriging was more suitable than the IDW or TIN methods for spatial interpolation of all parameters measured showing that transect data with irregular spatial distribution patterns are sensitive to interpolation methods.

Stanimirova *et al.* (1999) used Principal Component Analysis (PCA) on sediment data from the Gulf of Mexico to characterise various chemical components as 'inorganic natural', 'inorganic anthropogenic', 'bioorganic' and 'organic anthropogenic'. The method divides natural from anthropogenic influences and allows each participant in the sediment formation process to be used as marker of either natural or anthropogenic effects. Bell *et al.* (1999) examined the temporal dynamics of seagrass species distribution, their relationship to water depth and the serial replacement of one species by another in Tampa Bay, Florida. They used the Geographical Resources Analysis Support System (GRASS) GIS to examine the seagrass distributional change over a 2-year period, the seasonal and annual patterns of seagrass species transitions and the relation of these patterns to water depth. Gottgens (2000) used GIS to a 120-year record (1872–1991) of images of a 2000-hectare marsh system along the southwestern shore of Lake Erie, Ohio to record long-term variability in aggregate characteristics of wetland vegetation (e.g. environmental changes and human impact). Field and Philipp (2000) classified colour infrared aerial photography from the freshwater tidal wetlands of the Delaware River taken in 1978 and 1998. Marsh polygons were classified into either high-marsh or low-marsh, according to the dominant visual signature of the vegetation of each polygon and placed in a GIS for measuring the change in coverage areas of high- and low-marsh during that period. Cholwek *et al.* (2000) used a bottom classification sensor (RoxAnn®)

to map bottom surficial substrates in Minnesota's near shore waters of Lake Superior. Their main aim was to categorise spawning and nursery habitat for lake trout in the area. Delaney *et al.* (2000) compared five natural and 10 created *Spartina alterniflora* marshes in the Lower Galveston Bay System to determine the difference in physical characteristics associated with each type of marsh. They processed aerial photographs in a GIS to compare salt marshes on the basis of microhabitats, length/width ratio, area/perimeter ratio, marsh/water edge ratio, total size of plant communities, fetch distances, angle of exposure, orientation and elevation. Bodie and Semlitsch (2000) integrated telemetry data and classified aerial photography in a GIS to map habitat use by aquatic turtles. De Falco *et al.* (2000) studied the relationships between sediment distribution and *Posidonia oceanica* seagrass in the Gulf of Oristano (Sardinia, Italy) and found that sediment texture and composition is related to meadow spatial distribution. De Jong (2000) presented a GIS technique for the development of habitat maps by combining monoparametric maps that were classified. The classifications of these maps were based on the relations between the parameters and the habitat classes. Van der Merwe and Lohrentz (2001) used GIS as spatial decision support system to legitimise the demarcation of vegetated buffer zones in Saldanha Bay (South Africa). Their effort was based on conservation concepts through the protection of vegetated buffers and greenways.

A general description of the methodology used within GIS for the mapping of underwater habitats with the use of sonar and aerial photography data includes the following: Sonar point data are obtained during hydroacoustic surveys on-board research vessels while aerial photographs are acquired by an airplane, which flies at a certain altitude and produces coastal photographs of a certain scale. Sonar data are organised in a thematic coverage of point topology. Aerial photographs are registered and rectified. Depending on the clarity of the aerial data, photographs could either be automatically classified or used as base maps for on-screen digitising of polygons of different sediment types. Then, these polygons are either stored as classified sediment types or converted to points and placed with surveyed sonar point data in a common thematic coverage. In the first case, sonar data are interpolated (e.g. using kriging) and then converted to polygons, which are integrated with those polygons derived from the aerial photographs. In the second case, interpolation of the set of combined point values (from aerial photography and sonar data) provides a grid of substrate types, which then can be converted to polygons (see also Colour plate 3). The method of interpolation of sonar data plays an important role for the accuracy of the final bottom sediment map. Standardisation and harmonisation of the mapping, classification and GIS techniques used, as well as data storage are also essential in benthic habitat mapping.

## 2.12 SUMMARY

Applications of GIS in Oceanography are extremely diverse. GIS are used in almost any aspect of oceanographic research combining a variety of georeferenced data from on-board research surveys as well as from RS techniques. Together, EO data and the advancement of GIS technology bring new insights in oceanographic data processing. The major ocean processes may be observed through different satellite imagery through various parameters that characterise these processes and can be integrated in GIS. Data obtained through various mapping technologies (e.g. sonar, aerial photography, etc.) may be processed in GIS for mapping bathymetry and submerged aquatic vegetation.

Currently, the 3D nature of some of the ocean data (e.g. CDT) and the technological lack of 3D GIS databases constrict data integration into only two dimensions. A 3D ocean dataset may be only visualised using GIS and scientific visualisation packages, which offer some exciting capabilities in viewing the spatiotemporal development of several ocean processes through animated cartography. However, several marine GIS applications work around this problem by integrating 3D ocean data in multiple 2D layers, a practise often inadequate to appropriately model real world processes. The continuing expansion in both the amount of oceanographic data and GIS technology will definitely bring new innovative insights towards 3D GIS data integration.

Advances in technology and decreases in cost are making RS and GIS practical and attractive for use in marine resource management allowing researchers and managers to take a broader view of ecological patterns and processes. GIS processed RS oceanographic data provide quantitative estimates of marine, coastal and estuarine habitat conditions providing the means to monitor and manage Earth's global marine problems. This broad analytical monitoring of marine environment provides invaluable information for effective management efforts. Parallel advances in marine applications of GIS help incorporate ancillary data layers to improve the accuracy of raw satellite derived information. When these techniques for generating, organising, storing, analysing and integrating multisource spatial information are combined with mathematical models, coastal planners and ocean managers have a means for assessing the impacts of alternative management practices.

As marine researchers and managers take a broader, more holistic view of ecological patterns and marine processes, the synoptic and spatially referenced information over large marine regions provides a significant benefit of available and proposed RS and GIS techniques relevant to coastal and ocean management. Although in the past, the usefulness of RS and GIS has not reached its full potential, it now appears that need, cost and technology are converging in a way that continually proves practical and attractive through a diverse number of marine GIS studies, development of marine GIS tools for specific marine areas and production of several GIS-ready data products.

The Internet, in connection to oceanographic GIS applications, became a multilevel communication platform. Diffusion of oceanographic data through online publicly available data servers greatly facilitates the development of marine GIS databases. The number of online marine GIS tools is continually growing allowing the sharing of GIS processed information among management authorities. This is the most important benefit for environmental policy makers in their efforts to reach information-based decisions.

The use of GIS in Oceanography, as a relatively recently introduced application, will definitely expand in the future. One great development would be that of 3D databases capable of 3D data integration and technological efforts should focus on the creation of marine GIS packages with these capabilities.

## 2.13 REFERENCES

Al-Ghadban, A.N. (1997). Towards a GIS based Environmental Information System for Kuwait. In *Proceedings of the GIS/GPS Conference* 1997, Qatar. On line: http://www.gisqatar.org.qa/conf97/p1.html

Argo Science Team (1998). On the design and implementation of Argo: an initial plan for a global array of profiling floats. International CLIVAR Project Office Report 21, GODAE Report 5. GODAE International Project Office, Melbourne Australia, p. 32. On line: http://www.argo.ucsd.edu/oceanobs.html

Askari, F., Malaret, E., Lyzenga, D. and Collins, M. (1996). A PC based Remote Sensing system for detection and classification of oceanic fronts. In *Proceedings of the International Geoscience and Remote Sensing Symposium*, **2**, 1141–1145.

Aspinall, R. and Pearson, D. (2000). Integrated geographical assessment of environmental condition in water catchments: linking landscape ecology, environmental modelling and GIS. *Journal of Environmental Management*, **59**, 299–319.

Aswathanarayana, U. (1999). Functions and organisational structure of the proposed Natural Resources Management Facility in Mozambique. *Environmental Geology*, **37(3)**, 176–180.

Basu, A. (1998). Case study of land and marine data integration using GIS. *Surveying and Land Information Systems*, **58(3)**, 147–155.

Basu, A. and Saxena, N.K. (1999). A review of shallow water mapping systems. *Marine Geodesy*, **22(4)**, 249–257.

Belfiore, S. (2000). Recent developments in coastal management in the European Union. *Ocean and Coastal Management*, **43(1)**, 123–135.

Bell, S.S., Robbins, B.D. and Jensen, S.L. (1999). Gap dynamics in a seagrass landscape. *Ecosystems*, **2(6)**, 493–504.

Belokopytov, V.N. (1998). GIS hydrometeorology of the Black and Azov Seas. In *Proceedings of the International Symposium on Information Technology in Oceanography* (Ito 1998), October 1998, Goa India.

Berz, G., Kron, W., Loster, T., Rauch, E., Schimetschek, J., Schmieder, J., Siebert, A., Smolka, A. and Wirtz, A. (2001). World Map of Natural Hazards: A global view of the distribution and intensity of significant exposures. *Natural Hazards*, **23(2–3)**, 443–465.

Bettinetti, A., Pypaert, P. and Sweerts, J.P. (1996). Application of an integrated management approach to the restoration project of the Lagoon of Venice. *Journal of Environmental Management*, **46(3)**, 207–227.

Beusen, A.H.W., Klepper, O. and Meinardi, C.R. (1995). Modelling the flow of nitrogen and phosphorus in Europe: from loads to coastal seas. *Water Science and Technology*, **31(8)**, 141–145.

Bjorner, D. (1999). A triptych software development paradigm: domain, requirements and software towards a model development of a decision support system for sustainable development. *Lecture Notes in Computer Science*, **1710**, 29–60.

Blomgren, S. (1999). A digital elevation model for estimating flooding scenarios at the Falsterbo Peninsula. *Environmental Modelling and Software*, **14**, 579–587.

Bobbitt, A.M., Dziak, R.P., Stafford, K.M. and Fox, C.G. (1997). GIS analysis of oceanographic remotely sensed and field observation data. *Marine Geodesy*, **20(3–4)**, 153–161.

Bodie, J.R. and Semlitsch, R.D. (2000). Spatial and temporal use of floodplain habitats by lentic and lotic species of aquatic turtles. *Oecologia*, **122**, 138–146.

Brasseur, P., Brankart, J.M., Schoenauen, R. and Beckers, J.M. (1996). Seasonal temperature and salinity fields in the Mediterranean Sea: climatological analyses of an historical data set. *Deep Sea Research*, **43**, 159–192.

Bulthuis, D.A. (1995). Distribution of seagrasses in a North Puget Sound estuary: Padilla Bay, Washington, USA. *Aquatic Botany*, **50(1)**, 99–105.

Burrows, M. and Thorpe, S.A. (1999). Drifter observations of the Hebrides slope current and nearby circulation patterns. *Annales Geophysicae*, **17(2)**, 280–302.

Bushing, W.W. (1997). GIS based gap analysis of an existing marine reserve network around Santa Catalina Island. *Marine Geodesy*, **20**, 205–234.

Caldwell, P. (1999). Coral reef mapping: local partnerships in the Pacific support national effort: Geographic Information Systems help manage fragile coral reef ecosystem. *Earth System Monitor*, **9(4)**, 1–5.

Caloz, R. and Collet, C. (1997). Geographic information systems (GIS) and remote sensing in aquatic botany: methodological aspects. *Aquatic Botany*, **58**, 209–228.

Capobianco, M. (1999). Role and use of technologies in relation to ICZM. European Commission. On line: http://europa.eu.int/comm/dg11/iczm/themanal.htm

Cedfeldt, P.T., Watzin, M.C. and Richardson, B.D. (2000). Using GIS to identify functionally significant wetlands in the northeastern United States. *Environmental Management*, **26(1)**, 13–24.

Cholwek, G., Bonde, J., Li, X., Richards, C. and Yin, K. (2000). Processing RoxAnn sonar data to improve its categorization of lake bed surficial substrates. *Marine Geophysical Researches*, **21(5)**, 409–421.

Christensen, J.D., Monaco, M.E. and Lowery, T.E. (1997). An index to assess the sensitivity of Gulf of Mexico species to changes in estuarine salinity regimes. *Gulf Research Reports*, **9(4)**, 219–229.

Chua, T.E. (1997). The essential elements of science and management in coastal environmental managements. *Hydrobiologia*, **352(1–3)**, 159–166.

Clarke, J.E.H. (1999). Hardware and data transformations: a comparison of swath sonar systems, demonstrated at the 1996 US/Canada Hydrographic Commission Coastal Multibeam Surveying Course. On line: http://www.omg.unb.ca/~jhc/uschc96/comp_hardware.html

Cocker, M.D. and Shapiro, E.A. (1999). The Continental Margins Programme in Georgia. *Marine Georesources and Geotechnology*, **17(2)**, 199–209.

Condal, A. and Gold, C. (1995). A spatial data structure integrating GIS and simulation in a marine environment. *Marine Geodesy*, **18**, 213–228.

Crane, K., Galasso, J., Brown, C., Cherkashov, G., Ivanov, G. and Vanstain, B. (2000). Northern Ocean inventories of radionuclide contamination: GIS efforts to determine the past and present state of the environment in and adjacent to the Arctic. *Marine Pollution Bulletin*, **40(10)**, 853–868.

Crossney, S. (2000). Assessing the accuracy of LIDAR imagery for coastal management. In *Proceedings of 2000 ESRI International User Conference*. On line: http://www.esri.com/library/userconf/proc00/professional/abstracts/a765.htm

Crowley, M.F., Fracassi, J.F. and Glenn, S.M. (1999). Using real time remote sensing and *in situ* ocean data for adaptive sampling and data assimilative modelling. In *Proceedings of The American Society of Limnology and Oceanography, Aquatic Sciences Meeting: Limnology and Oceanography: Navigating into the Next Century*, Santa Fe, New Mexico, February 1999, http://www.aslo.org/santafe99/

Cummings, J.A. (1994). Global and regional ocean thermal analysis systems at Fleet Numerical Meteorology and Oceanography Centre. *IEEE Transactions on Geoscience and Remote Sensing*, **32(6)**, 234–245.

Cutter, G.R., Diaz, R.J. and Hobbs, C.H. (1998). Benthic habitats and biological resources off the Virginia coast 1996 and 1997. Report for MMS Cooperative Agreement 14/35/0001/3087. On line: http://www.vims.edu/~cutter/vabeach/vbtitle.html.

Davies, J., Foster–Smith, R. and Sotheran, I.S. (1997). Marine biological mapping for environment management using acoustic ground discrimination systems and Geographic Information Systems. *Underwater Technology*, **22(4)**, 167–172.

Davis, H.H., Caldwell, P.D., Goodwin, P.B. and Karver, E. (1994). Use of SPOT imagery to obtain GIS input for oil spill models. *Earth Observations Magazine*, **3(12)**, 30–33.

Davis-Lunde, K., Jugan, L. and Shoemaker, J.T. (1999). Use of LOGIC to support lidar operations, *Proceedings of SPIE, The International Society for Optical Engineering*, **3761**, 29–33.

De Falco, G., Ferrari, S., Cancemi, G. and Baroli, M. (2000). Relationship between sediment distribution and *Posidonia oceanica* seagrass. *Geo-Marine Letters*, **20**, 50–57.

De Jong, D.J. (2000). Ecotopes in the Dutch marine tidal waters: A proposal for a classification of ecotopes and a method to map them. In *Proceedings of ICES Annual Science Conference 2000*, Theme Session on Marine Habitat Classification and Mapping, T:05, September 2000, Brugge, Belgium.

De Jonge, V.N., De Jong, D.J. and Van Katwijk, M.M. (2000). Policy plans and management measures to restore eelgrass (*Zostera marina* L.) in the Dutch Wadden Sea. *Helgoland Marine Research*, **54**,151–158.

Delaney, T.P., Webb, J.W. Jr and Minello, T.J. (2000). Comparison of physical characteristics between created and natural marshes in Galveston Bay, Texas. *Wetlands Ecology and Management*, **8**, 342–352.

De Lauro, M., Giumta, G. and Montella, R. (1999). Marine GIS Development: Mapping the Bay of Napes. *Sea Technology*, **40(6)**, 53–59.

DeAngelis, D.L., Gross, L.J., Huston, M.A., Wolff, W.F., Fleming, D.M., Comiskey, E.J. and Sylvester, S.M. (1998). Landscape Modelling for Everglades Ecosystem Restoration. *Ecosystems*, **1**, 64–75.

Delarue, A., Smith, S. and Edgar, A. (1999). AUV data processing and visualisation using GIS and Internet techniques. *IEEE Oceans Conference Record*, **2**, 738–742.

Demarcq, H. and Faure, V. (2000). Coastal upwelling and associated retention indices derived from satellite SST. Application to *Octopus vulgaris* recruitment. *Oceanologica Acta*, **23(4)**, 391–408.

De Oliveira, J.L., Pires, F. and Medeiros, C.B. (1997). An environment for modelling and design of geographic applications. *Geoinformatica*, **1(1)**, 29–58.

Dijkstra, H.A. and Molemaker, M.J. (1999). Imperfections of the North Atlantic wind driven ocean circulation: continental geometry and wind stress shape. *Journal of Marine Research*, **57(1)**, 1–28.

Dobosiewicz, J. (2001). Applications of digital elevation models and geographic information systems to coastal flood studies along the shoreline of Raritan Bay, New Jersey. *Environmental Geosciences*, **8(1)**, 11–20.

Donoghue, D.N.M., Thomas, D.C.R. and Zong, Y. (1994). Mapping and monitoring the intertidal zone of the east coast of England using remote sensing techniques and a coastal monitoring GIS. *Marine Technology Society Journal*, **28(2)**, 19–29.

Douligeris, C., Collins, J., Iakovou, E., Sun, P., Riggs, R. and Mooers, C.N.K. (1995). Development of OSIMS: an oil spill information management system. *Spill Science and Technology Bulletin*, **2(4)**, 255–263.

Drakopoulos, P.G., Valavanis, V. and Georgakarakos, S. (2000). Spatial and temporal distribution of chlorophyll in the Aegean Sea according to SeaWiFS imagery. In *Proceedings of the 6th Hellenic Symposium on Oceanography and Fisheries*, May 2000, Chios, Greece.

Dubuisson-Jolly, M.P. and Gupta, A. (2000). Color and texture fusion: application to aerial image segmentation and GIS updating. *Image and Vision Computing*, **18**, 823–832.

Ducrotoy, J.P., Elliott, M. and de Jonge, V.N. (2000). The North Sea. *Marine Pollution Bulletin*, **41(1–6)**, 5–23.

Durand, C. (1996). Presentation de Patelle: Programmemation graphique de chaines de traitement SIG sous ArcView. IFREMER Note Technique TC/04. Plouzane: Sillage.

Edgar, G.J., Barrett, N.S., Graddon, D.J. and Last, P.R. (2000). The conservation significance of estuaries: a classification of Tasmanian estuaries using ecological, physical and demographic attributes as a case study. *Biological Conservation*, **92(3)**, 383–397.

El-Raey, M. (1997). Vulnerability assessment of the coastal zone of the Nile delta of Egypt, to the impacts of sea level rise. *Ocean and Coastal Management*, **37(1)**, 29–40.

El-Raey, M., Fouda, Y. and Nasr, S. (1997). GIS assessment of the vulnerability of the Rosetta area, Egypt to impacts of sea rise. *Journal of Environmental Monitoring and Assessment*, **47**, 59–77.

Ferguson, R.L. and Korfmacher, K.F. (1997). Remote sensing and GIS analysis of seagrass meadows in North Carolina, USA. *Aquatic Botany*, **58**, 241–258.

Field, R.T. and Philipp, K.R. (2000). Vegetation changes in the freshwater tidal marsh of the Delaware estuary. *Wetlands Ecology and Management*, **8**, 79–88.

Fowler, C. and Gore, J. (1997). Creating a GIS for ocean planning and governance. In *Proceedings of 1997 ESRI International User Conference*. On line: http://www.esri.com/library/userconf/proc97/PROC97/TO550/PAP512/P512.HTM

Fowler, C. and Schmidt, N. (1998). Geographic information systems, mapping, and spatial data for the coastal and ocean resource management community. *Surveying and Land Information Systems*, **58(3)**, 135–140.

Fox, C.G. and Bobbit, A.M. (2000). The National Oceanic and Atmospheric Administration's Vents Programme GIS: Integration, Analysis, and Distribution of Multidisciplinary Oceanographic Data. In D. Wright and D. Bartlett, eds. *Marine and Coastal Information Systems*, pp. 163–176. Taylor and Francis, USA and Canada.

Ganas, A. and Papoulia, I. (2000). High resolution, digital mapping of the seismic hazard within the Gulf of Evia Rift, Central Greece, using normal fault segments as line sources. *Natural Hazards*, **22**, 203–223.

Gao, J. and O'Leary, S.M. (1997). Estimation of suspended solids from aerial photographs in a GIS. *International Journal of Remote Sensing*, **18**, 2073–2086.

Garcia, G.M., Pollard, J. and Rodriguez, R.D. (2000). Origins, management, and measurement of stress on the coast of Southern Spain. *Coastal Management*, **28(3)**, 215–234.

Garrabou, J. (1998). Applying a Geographical Information System (GIS) to the study of the growth of benthic clonal organisms. *Marine Ecology Progress Series*, **173**, 227–235.

Goldfinger, C. (2000). Active tectonics: data acquisition and analysis with marine GIS. In D. Wright and D. Bartlett, eds. *Marine and Coastal Information Systems*, pp. 237–254. Taylor and Francis, USA and Canada.

Goldfinger, C. and McNeill, L.C. (1997). Case study of GIS data integration and visualisation in submarine tectonic investigations: Cascadia subduction zone. *Marine Geodesy*, **20**, 267–289.

Gomis, D., Ruiz, S. and Pedder, M.A. (2001). Diagnostic analysis of the 3D ageostrophic circulation from a multivariate spatial interpolation of CTD and ADCP data. *Deep Sea Research I*, **48**, 269–295.

Gorman, L., Morang, A. and Larson, R. (1998). Monitoring the coastal environment; part IV: Mapping, shoreline changes, and bathymetric analysis. *Journal of Coastal Research*, **14(1)**, 61–92.

Gottgens, J.F. (2000). Wetland restoration along the southwestern Lake Erie coastline: case studies and recommendations. *Toledo Journal of Great Lakes Law, Science and Policy*, **3(1)**, 49–65.

Greve, C.A., Cowell, P.J. and Thom, B.G. (2000). Application of a Geographical Information System for risk assessment on open ocean beaches: Collaroy/ Narrabeen Beach, Sydney, Australia: an example. *Environmental Geosciences*, **7(3)**, 149–161.

Guan, W., Chamberlain, R.H., Sabol, B.M. and Doering, P.H. (1999). Mapping submerged aquatic vegetation with GIS in the Caloosahatchee Estuary: evaluation of different interpolation methods. *Marine Geodesy*, **22(2)**, 69–91.

Haltuch, M.A. and Berkman, P.A. (1999). Modelling expansion of exotic mussels on Lake Erie sediments using Geographic Information Systems. In *Proceedings of The American Society of Limnology and Oceanography, Aquatic Sciences Meeting: Limnology and Oceanography: Navigating into the Next Century*, February 1999, Santa Fe New Mexico. On line: http://www.aslo.org/santafe99/

Harrison, W.G., Aristegui, J., Head, E.J.H., Li, W.K.W., Longhurst, A.R. and Sameoto, D.D. (2001). Basin scale variability in plankton biomass and community metabolism in the sub tropical North Atlantic Ocean. *Deep Sea Research II*, **48**, 2241–2269.

Hatcher, G.A. and Maher, N.M. (2000). Real time GIS for marine applications. In D. Wright and D. Bartlett, eds. *Marine and Coastal Geographical Information Systems*, pp. 137–147. Taylor and Francis, USA and Canada.

Hatcher, G.A., Maher, N.M. and Orange, D.L. (1997). The customisation of ArcView as a real time tool for oceanographic research. In *Proceedings of the 1997 ESRI User Conference*, San Diego CA.

Hennecke, W.G., Greve, C.A. and Cowell, P.J. (2000). GIS modelling of potential impacts of near future sea level rise for managing coastal areas in southeastern Australia. In *Proceedings of the 4th International Conference on Integrating Geographic Information Systems (GIS) and Environmental Modelling*, September 2000, Banff, Alberta, Canada.

Hesselmans, G.H.F.M., Wensink, G.J. and Calkoen, C.J. (1997). Possibilities of remote sensing technologies in coastal studies. *GeoJournal*, **42(1)**, 65–72.

Hill, D., Coquillard, P., De Vaugelas, J. and Meinesz, A. (1998). An algorithmic model for invasive species: application to *Caulerpa taxifolia* (Vahl) C. Agardh development in the northwestern Mediterranean Sea. *Ecological Modelling*, **109(3)**, 251–265.

Hinton, J.C. (1996). GIS and Remote Sensing Integration for Environmental Applications. *International Journal of Geographical Information Science*, **10(7)**, 877–890.

Holasek, R.E., Portigal, F.P., Mooradian, G.C., Voelker, M.A., Even, D.M., Fene, M.W., Owensby, P.D. and Breitwieser, D.S. (1997). HSI mapping of marine and coastal environments using the advanced airborne hyperspectral imaging system (AAHIS). *The International Society for Optical Engineering*, **3071**, 169–180.

Hooge, P.N., Hooge, R.H., Solomon, E.K., Dezan, C.L., Dick, C.A. and Mondragon, J. (2000). Fjord Oceanography Monitoring Handbook: Glacier Bay, Alaska. On line: http://www.absc.usgs.gov/glba/oceanography/PROTOCOL/oceanography_handbook.pdf

Huettmann, F. and Diamond, A.W. (2001). Seabird colony locations and environmental determination of seabird distribution: a spatially explicit breeding seabird model for the Northwest Atlantic. *Ecological Modelling*, **141(1–3)**, 261–298.

IOC (1996a). GTSPP (Global Temperature/Salinity Pilot Project): Improving the availability of temperature and salinity data. Intergovernmental Oceanographic Commission (IOC) and World Meteorological Organisation (WMO). On line: http://www.incoir.org/gtspp.htm

IOC (1996b). Intergovernmental Oceanographic Commission of Unesco. Joint IOC/WMO Steering Group on Global Temperature-Salinity Pilot Project. Fourth Session, Washington DC, USA, April 1996. Summary Report. On line: http://www.meds-sdmm.dfo-mpo.gc.ca/ALPHAPRO/gtspp/meetings/wash96.htm

Islam, M.M. and Sado, K. (2000). Flood hazard assessment in Bangladesh using NOAA AVHRR data with Geographical Information System. *Hydrological Processes*, **14(3)**, 605–620.

Jansson, A., Folke, C. and Langaas, S. (1998). Quantification of the nitrogen retention capacity in natural wetlands in the Baltic drainage basin. *Landscape Ecololy*, **13**, 249–262.

Jansson, A., Folke, C., Rockstrom, J. and Gordon, L. (1999). Linking freshwater flows and ecosystem services appropriated by people: the case of the Baltic Sea drainage basin. *Ecosystems*, **2**, 351–366.

Jingsong, M. and Ying, W. (1999). A spatiotemporal data model on relational databases for coastal dynamic research. *Marine Geodesy*, **22(2)**, 105–114.

Johannessen, O.M., Pettersson, L.H., Bjorgo, E., Espedal, H., Evensen, G., Hamre, T., Jenkins, A., Korsbakken, E., Samuel, P. and Sandven, S. (1997). A review of the possible applications of earth observation data within EuroGOOS. In J. Stel, H.W.A. Behrens, J.C. Borst, L.J. Droppert and J.P. van der Meulen, eds.

*Operational Oceanography: The Challenge for European Cooperation*, pp. 192–205. Elsevier.

Johannessen, O.M., Sandven, S., Jenkins, A.D., Durand, D., Pettersson, L.H., Espedal, H., Eversen, G. and Hamre, T. (2000). Satellite earth observation in operational oceanography. *Coastal Engineering*, **41**, 155–176.

Johnson, A.K.L., Ebert, S.P. and Murray, A.E. (1999). Distribution of coastal freshwater wetlands and riparian forests in the Herbert River catchment and implications for management of catchments adjacent the Great Barrier Reef Marine Park. *Environmental Conservation*, **26(3)**, 229–235.

Jones, R. (2000). Using GIS and the sea level affecting marshes model (SLAMM4) to model shorebird habitat loss due to global climate change. In *Proceedings of 2000 ESRI International User Conference*. On line: http://www.esri.com/library/userconf/proc00/professional/abstracts/a160.htm

Karl, D.M. (1999). A sea of change: biogeochemical variability in the North Pacific Subtropical Gyre. *Ecosystems*, **2**, 181–214.

Khedouri, E., Gemmill, W. and Shank, M. (1976). Statistical summary of ocean fronts and water masses in the Western North Atlantic. USA Naval Oceanographic Office, Report NOO RP–9.

King, I. (2000). CPACC: A regional approach to the application of GIS for adaptation planning to global climate change and sea level rise. In *Proceedings of 2000 ESRI International User Conference*. On line: http://www.esri.com/library/userconf/proc00/professional/papers/PAP333/p333.htm

Kitamoto, A. and Takagi, M. (1999). Image classification using probabilistic models that reflect the internal structure of mixels. *Pattern Analysis and Applications*, **2(1)**, 31–43.

Kitsiou, D. and Karydis, M. (2000). Categorical mapping of marine eutrophication based on ecological indices. *The Science of The Total Environment*, **255(1–3)**, 113–127.

Klemas, V.V. (2001). Remote sensing of landscape level coastal environmental indicators. *Environmental Management*, **27(1)**, 47–57.

Kleypas, J.A., Buddemeier, R.W. and Gattuso, J.P. (2000). Defining coral reef for the age of global change. *International Journal of Earth Science*, Springer Link, Online First: http://link.springer.de/

Knudsen, T. (1999). Busstop: an integrated system for handling, analysis, and visualisation of ocean data. *Physics and Chemistry of the Earch (A)*, **24**, 411–414.

Krishnan, P. (1995). A geographical information system for oil spills sensitivity mapping in the Shetland Islands (United Kingdom). *Ocean and Coastal Management*, **26(3)**, 247–255.

Kruer, C. (1995). Mapping and characterizing seagrass areas important to manatees in Puerto Rico: benthic communities mapping and assessment. Report to US DoI, National Biological Service, Sirenia Project, p. 14, No. 83023/5/0161.

Kunte, P.D. (1995). Worldwide databases in marine geology: A review. *Marine Geology*, **122(3)**, 263–275.

Lathrop, R.G., Cole, M.B. and Showalter, R.D. (2000). Quantifying the habitat structure and spatial pattern of New Jersey (USA) salt marshes under different management regimes. *Wetlands Ecology and Management*, **8(2–3)**, 163–172.

Lee, H.J., Locat, J., Dartnell, P. and Israel, K. (1999). Regional variability of slope stability: application to the Eel Margin, California. *Marine Geology*, **154**, 305–322.

Lehmann, A., Jaquet, J.M. and Lachavanne, J.B. (1994). Contribution of GIS to submerged macrophyte biomass estimation and community structure modelling, Lake Geneva, Switzerland. *Aquatic Botany*, **47**, 99–117.

Leshkevich, G.A. and Liu, S. (2000). Internet access to Great Lakes CoastWatch remote sensing information. *International Geoscience and Remote Sensing Symposium (IGARSS)*, **5**, 2077–2079.

Li, G. and Shao, Y. (1998). Remote sensing of oceanic primary productivity and its GIS estimation model. *Acta Geographica Sinica*, **53(6)**, 546–553.

Li, R., Keong, C.W., Ramcharan, E., Kjerfve, B. and Willis, D. (1998). A Coastal GIS for Shoreline Monitoring and Management: Case Study in Malaysia. *Surveying and Land Information Systems*, **58(3)**, 157–166.

Li, R., Liu, J.K. and Felus, Y. (2001). Spatial modelling and analysis for shoreline change detection and coastal erosion monitoring. *Marine Geodesy*, **24(1)**, 1–12.

Li, Y., Brimicombe, A.J. and Ralphs, M.P. (2000). Spatial data quality and sensitivity analysis in GIS and environmental modelling: the case of coastal oil spills. *Computers, Environment and Urban Systems*, **24**, 95–108.

Lienert, B.R., Porter, J.N. and Sharma, S.K. (1999). Real time analysis and display of scanning Lidar scattering data. *Marine Geodesy*, **22(4)**, 259–265.

Lin, H., Lu, G., Song, Z. and Gong, J. (1999). Modelling the tide system of the East China Sea with GIS. *Marine Geodesy*, **22(2)**, 115–128.

Liu, H., Jezek, K.C. and Li, B. (1999). Development of an Antarctic digital elevation model by integrating cartographic and remotely sensed data: A geographic information system based approach. *Journal of Geophysical Research B: Solid Earth*, **104**, 23199–23213.

Livingstone, D., Raper, J. and McCarthy, T. (1999). Integrating aerial videography and digital photography with terrain modelling: an application for coastal geomorphology. *Geomorphology*, **29(1–2)**, 77–92.

Lybanon, M. (1996). Maltese Front variability from satellite observations based on automated detection. *IEEE Transactions on Geoscience and Remote Sensing*, **34(5)**, 1159–1165.

MacDonald, A. and Cain, M. (2000). Marine environmental high risk areas (MEHRAs) for the UK. *International Maritime Technology*, **112(2)**, 61–70.

Mallinson, D., Hine, A., Naar, D., Hafen, M., Schock, S., Smith, S., Gelfenbaum, G., Wilson, D. and Lavoie, D. (1997). Seafloor mapping using the Autonomous Underwater Vehicle (AUV) Ocean Explorer. *EOS, Transactions of the American Geophysical Union, Fall Supplement*, **78**, F350.

Marchisio, G.B., Koperski, K. and Sanella, M. (2000). Querying remote sensing and GIS repositories with spatial association rules. *International Geoscience and Remote Sensing Symposium (IGARSS)*, **7**, 3054–3057.

May, D.A. (1993). Global and regional comparative performance of linear and nonlinear satellite multichannel sea surface temperature algorithms. Remote Sensing Division, USA Naval Research Laboratory, NRL/MR/7240—93–7049.

McAdoo, B.G., Pratson, L.F. and Orange, D.L. (2000). Submarine landslide geomorphology, US continental slope. *Marine Geology*, **169(1–2)**, 103–136.

McRea, Jr J.E., Greene, H.G., O'Connell, V.M. and Wakefield, W.W. (1999). Mapping marine habitats with high resolution sidescan sonar. *Oceanologica Acta*, **22(6)**, 679–686.

Menon, H.B. (1998). Role of oceanic fronts in promoting productivity in the southern Indian ocean. In K. Sherman, E. Okemwa and M. Ntiba, eds. *Large Marine Ecosystems of the Indian Ocean: Assessment, Sustainability, and Management*, p. 394. Blackwell Science, Cambridge.

Mesick, S., Hudson, M. and Gathof, J. (2000). Ocean survey planning and tracking: integrating technology to support naval operations. In *Proceedings of 2000 ESRI International User Conference*. On line: http://www.esri.com/library/userconf/proc00/professional/papers/PAP623/p623.htm

Mesick, S.M., Booda, M.H. and Gibson, B.A. (1998). Automated detection of oceanic fronts and eddies from remotely sensed satellite data using ARC/INFO GRID processing. In *Proceedings of 1998 ESRI International User Conference*.

On line: http://www.esri.com/library/userconf/proc98/PROCEED/TO300/PAP252/P252.HTM

Moe, K.A., Skeie, G.M., Brude, O.W., Lovas, S.M., Nedrebo, M. and Weslawski, J.M. (2000). The Svalbard intertidal zone: a concept for the use of GIS in applied oil sensitivity, vulnerability and impact analyses. *Spill Science and Technology Bulletin*, **6(2)**, 187–206.

Moore, J.K., Abbott, M.R. and Richman, J.G. (1999). Location and dynamics of the Antarctic Polar Front from satellite sea surface temperature data. *Journal of Geophysical Research*, **104**, 3059–3073.

Mumby, P.J., Green, E.P., Edwards, A.J. and Clark, C.D. (1999). The cost effectiveness of remote sensing for tropical coastal resources assessment and management. *Journal of Environmental Management*, **55**, 157–166.

Muskat, J. (2000). GIS applications for oil spill prevention and response in California. In *Proceedings of 2000 ESRI International User Conference*. On line: http://www.esri.com/library/userconf/proc00/professional/papers/PAP786/p786.htm

Napolitano, E., Oguz, T., Malanotte-Rizzoli, P., Yilmaz, A. and Sansone, E. (2000). Simulations of biological production in the Rhodes and Ionian basins of the eastern Mediterranean. *Journal of Marine Systems*, **24(3–4)**, 277–298.

Nasr, S., El-Raey, M., Ezzat, H. and Ibrahim, A. (1997). Geographical information system analysis for sediments heavy metals and pesticides in Abu Qir Bay, Egypt. *Journal of Coastal Research*, **13(4)**, 1233–1237.

Neilson, B. and Costello, M.J. (1999). The Relative lengths of seashore substrata around the coastline of Ireland as determined by digital methods in a Geographical Information System. *Estuarine, Coastal and Shelf Science*, **49(4)**, 501–508.

NetWatch Science Magazine (2000). EDUCATION: Lab on the Lake. *Science*, **290(5494)**, 1047.

Nico, G., Pappalepore, M., Pasquariello, G., Refice, A. and Samarelli, S. (2000). Comparison of SAR amplitude vs. coherence flood detection methods: a GIS application. *International Journal of Remote Sensing*, **21(8)**, 1619–1631.

Noji, T., Thorsnes, T. and Fossa, J.H. (2000). Marine habitat mapping for the Norwegian Sea. ICES 2000 Annual Science Conference 27–30 September 2000, 88th Statutory Meeting, 24 September to 4 October 2000, Brugge, Belgium. ICES CM 2000/T:13.

O'Driscoll, R.L. and McClatchie, S. (1998). Spatial distribution of planktivorous fish schools in relation to krill abundance and local hydrography off Otago, New Zealand. *Deep Sea Research II*, **45**, 1295–1325.

Olsvig-Whittaker, L., Friedman, D., Magal, Y. and Shurkey, R. (2000). Conservation management modelling at En Afeq Nature Reserve, Israel. *Proceedings of the Yearly IAVS Symposium*, **1998(1)**, 294–297.

Ong'anda, H.O. (1997). Marine environmental and oceanographic data management. In *Proceedings of the 1st Western Indian Ocean Marine Science Association (WIOMSA): Scientific Symposium on Advances in Marine Science in Eastern Africa*, May 1997, Mombasa Kenya.

Passi, R.M. and Harsh, A. (1994). Objective feature identification and tracking: a review. Mississippi State Centre for Air and Sea Technology, Technical Report 94-4.

Pasqualini, V., Pergent-Martini, C., Clabaut, P. and Pergent, G. (1998). Mapping of *Posidonia oceanica* using aerial photographs and side scan sonar: application off the Island of Corsica (France). *Estuarine, Coastal and Shelf Science*, **47**, 359–367.

Paton, M.H., Dietrich, K., Liew, O., Dinn, A. and Patrick, A. (2000). VESPA: a benchmark for vector spatial databases. *Lecture Notes in Computer Science*, **1832**, 81–101.

Plewe, B. (1997). *GIS Online: Information Retrieval, Mapping and the Internet*, p. 311. OnWord Press, Albany, New York.

Populus, J., Moreau, F., Coquelet, D. and Xavier, J.P. (1995). An assessment of environmental sensitivity to marine pollutions: solutions with remote sensing and geographical information systems (GIS). *International Journal of Remote Sensing*, **16(1)**, 3–15.

Pratson, L., Divins, D., Butler, T., Metzger, D., Steele, M., Sharman, G., Berggren, T., Holcombe, T. and Ramos, R. (1999). Development of an elevation database for the US coastal zone. *Surveying and Land Information Systems*, **59(1)**, 3–13.

Rea, T.E., Karapatakis, D.J., Guy, K.K., Pinder, J.E. and Mackey, H.E. (1998). The relative effects of water depth, fetch and other physical factors on the development of macrophytes in a small southeastern US pond. *Aquatic Botany*, **61**, 289–299.

Rymell, M.C., Sabeur, Z.A., Williams, M.O., Borwell, D.M. and Tyler, A.O. (1997). Development and application of environmental models in the assessment of exploratory drilling in a sensitive coastal region, Isle of Man. In *Proceedings of the UK, SPE/UKOOA European Environmental Conference*, 1997, pp. 257–265.

Schmieder, K. (1997). Littoral zone GIS of Lake Constance: a useful tool in lake monitoring and autecological studies with submersed macrophytes. *Aquatic Botany*, **58(3–4)**, 333–346.

Scholten, H.J., LoCashio, A. and Overduin, T. (1998). Towards a spatial information infrastructure for flood management in the Netherlands. *Journal of Coastal Conservation*, **4(2)**, 151–160.

Semlitsch, R.D. and Bodie, J.R. (1998). Are small, isolated wetlands expendable? *Conservation Biology*, **12(5)**, 1129–1134.

Shafer, C.S. and Benzaken, D. (1998). User perceptions about marine wilderness on Australia's Great Barrier Reef. *Coastal Management*, **26(2)**, 79–91.

Sharma, R., Gopalan, A.K.S. and Ali, M.M. (1999). Interannual variation of eddy kinetic energy from TOPEX altimeter observations. *Marine Geodesy*, **22(4)**, 239–248.

Shaw, A.G.P. and Vennell, R. (2001). Measurements of an oceanic front using a front following algorithm for AVHRR SST imagery. *Remote Sensing of Environment*, **75**, 47–62.

Siakavara, K., Valavanis, V. and Banks, A.C. (2000). Mapping of NATURA 2000 marine benthic habitats using remote sensing techniques. In *Proceedings of the 1st Mediterranean Symposium on Marine Vegetation*, Regional Activity Centre for Specially Protected Areas (RAC/SPA), October 2000, Ajaccio, Corsica.

Sklar, F.H. and Browder, J.A. (1998). Coastal environmental impacts brought about by alterations to freshwater flow in the Gulf of Mexico. *Environmental Management*, **22(4)**, 547–562.

Simas, T., Nunes, J.P. and Ferreira, J.G. (2001). Effects of global climate change on coastal salt marshes. *Ecological Modelling*, **139**, 1–15.

Smith, H.D. and Lalwani, C.S. (1996). The North Sea: coordinated sea use management. *GeoJournal*, **39(2)**, 109–115.

Smith, L.A. and Loza, L. (1994). Texas turns to GIS for oil spill management. *Geo Info Systems*, **4(2)**, 48–50.

Smith, W.H.F. and Sandwell, D.T. (1997). Global sea floor topography from satellite altimetry and ship depth soundings. *Science*, **277**, 1956–1962.

Smith, R.A. and West, G.R. (1999). Airborne LIDAR: A surveying tool for the new millennium. In *Proceedings of Oceans '99 MTS/IEEE*, September 1999, Seattle, WA.

Solanki, H.U., Dwivedi, R.M. and Narain, A. (1998). Satellite obseravtions of coastal upwelling in the North Arabian Sea and its impacts on the fishery resources. In *Proceedings of the International Symposium on Information Technology in Oceanography* (Ito 1998), October 1998, Goa, India, pp. 12–16.

Sorensen, M. (1995). ARC/INFO marine spill GIS. *Spill Science and Technology Bulletin*, **2(1)**, 81–85.

Sowmya, A. and Trinder, J. (2000). Modelling and representation issues in automated feature extraction from aerial and satellite images. *ISPRS Journal of Photogrammetry and Remote Sensing*, **55**, 34–47.

Spalding, M.D. and Grenfell, A.M. (1997). New estimates of global and regional coral reef areas. *Coral Reefs*, **16**, 225–230.

Spaulding, M.L., Mendelsohn, D.L. and Swanson, J.C. (1999). WQMAP: an integrated three dimensional hydrodynamic and water quality model system for estuarine and coastal applications. *Marine Technology Society Journal*, **33(3)**, 38–54.

Stanbury, K.B. and Starr, R.M. (1999). Applications of Geographic Information Systems (GIS) to habitat assessment and marine resource management. *Oceanologica Acta*, **22(6)**, 699–703.

Stanimirova, I., Tsakovski, S., Simeonov, V. (1999). Multivariate statistical analysis of coastal sediment data. *Fresenius' Journal of Analytical Chemistry*, **365(6)**, 489–493.

Stejskal, I.V. (2000). Obtaining approvals for oil and gas projects in shallow water marine areas in Western Australia using an environmental risk assessment framework. *Spill Science and Technology Bulletin*, **6(1)**, 69–76.

Su, Y. and Sheng, Y. (1999). Visualizing upwelling at Monterey Bay in an integrated environment of GIS and scientific visualisation. *Marine Geodesy*, **22(2)**, 93–104.

Su, Y., Slottow, J. and Mozes, A. (2000). Distributing proprietary geographic data on the World Wide Web: UCLA GIS Database and Map Server. *Computers and Geosciences*, **26(7)**, 741–749.

Suzanne, M. and Lybanon, M. (1983). Automated boundary delineation in infrared ocean images. *IEEE Transaction on Geoscience and Remote Sensing*, **31(6)**, 123–134.

Thia-Eng, C. (1999). Marine pollution prevention and management in the East Asian Seas: a paradigm shift in concept, approach and methodology. *Marine Pollution Bulletin*, **39(1–9)**, 80–88.

Thorpe, S.A. (1998). Turbulence in the stratified and rotating World Ocean. *Theoretical and Computational Fluid Dynamics*, **11**, 171–181.

Thumerer, T., Jones, A.P. and Brown, D. (2000). A GIS based coastal management system for climate change associated flood risk assessment on the east coast of England. *International Journal of GIS*, **14(3)**, 265–281.

Titus, J.G. and Richman, C. (2001). Maps of lands vulnerable to sea level rise: modeled elevations along the US Atlantic and Gulf Coasts. *Climate Research*, **16(4)**, 275–282.

Tortell, P. and Awosika, L. (1996). Oceanographic survey techniques and living resources assessment methods. Intergovernmental Oceanographic Commission, Manuals and Guides No. 32, UNESCO 1996. Online: http://ioc.unesco.org/iocweb/iocpub/iocpdf/m032.pdf

Tsanis, I.K. and Boyle, S. (2001). A 2D hydrodynamic/pollutant transport GIS model. *Advances in Engineering Software*, **32**, 353–361.

Tsurumi, M. (1998). The application of Geographical Information Systems to biological studies at hydrothermal vents. *Cahiers de Biologie Marine*, **39(3–4)**, 263–266.

UN (1997). Programme for the Environmental Management and Protection of the Black Sea 1997: Black Sea GIS. United Nations Publications, New York.

Valavanis, V.D., Drakopoulos, P. and Georgakarakos, S. (1999). A Study of upwellings using GIS. In *Proceedings of CoastGIS 1999 International Conference on GIS and New Advances in Integrated Coastal Management*, September 1999, Brest, France.

Valavanis, V.D., Drakopoulos, P. and Georgakarakos, S. (2000). Upwelling identification and measurement system. In *Proceedings of the 10th Hellenic Geographic Symposium*, October 2000, Athens, Greece.

Valavanis, V.D., Georgakarakos, S. and Haralabus, J. (1998). A methodology for GIS interfacing of marine data. In *Proccedings of GISPlaNET 98 International Conference and Exhibition of Geographic Information*, September 1998, Lisbon, Portugal.

Van der Merwe, J.H. and Lohrentz, G. (2001). Demarcating coastal vegetation buffers with multicriteria evaluation and GIS at Saldanha Bay, South Africa. *Ambio*, **30(2)**, 89–95.

Van Zuidam, R.A., Farifteh, J., Eleveld, M.A. and Cheng, T. (1998). Developments in remote sensing, dynamic modelling and GIS applications for integrated coastal zone management. *Journal of Coastal Conservation*, **4(2)**, 191–202.

Wai, W.Y.K. (1994). A computational model for detecting image changes. Master's Thesis, University of Toronto.

Waluda, C.M., Rodhouse, P.G., Trathan, P.N. and Pierce, G.J. (2001). Remotely sensed mesoscale oceanography and the distribution of *Illex argentinus* in the South Atlantic. *Fisheries Oceanography*, **10(2)**, 207–216.

Wang, X. (2001). Integrating water quality management and land use planning in a watershed context. *Journal of Environmental Management*, **61(1)**, 25–36.

Wessel, P. and Smith, W.H.F. (1995). New version of the Generic Mapping Tools released. *EOS Transactions American Geophysical Union*, **76(33)**, 329.

West, L.A. (1999). Florida's Marine Resource Information System: A geographic decision support system. *Government Information Quartely*, **16(1)**, 47–62.

Wilkinson, G.G. (1996). A Review of current issues in the integration of GIS and Remote Sensing data. *International Journal of Geographical Information Science*, **10(1)**, 85–101.

Wong, F.L., Eittreim, S.L., Degnan, C.H. and Lee, W.C. (1999). USGS seafloor GIS for Monterey Sanctuary: selected data types. In *Proceedings of 1999 ESRI International User Conference*, San Diego, California.

Wright, D.J. (1996). Rumblings on the ocean floor: GIS supports deepsea research. *Geographical Information Systems*, **6**, 22–29.

Wright, D.J. (1999). Getting to the bottom of it: Tools, techniques, and discoveries of deep ocean geography. *The Professional Geographer*, **51(3)**, 426–439.

Wright, D.J., Bloomer, S.H., MacLeod, C.J., Taylor, B. and Goodliffe, A.M. (2000). Bathymetry of the Tonga Trench and forearc: A map series. *Marine Geophysical Researches*, **21(5)**, 489–512.

Wright, D.J., Fox, C.G. and Bobbitt, A.M. (1997). A scientific information model for deepsea mapping and sampling. *Marine Geodesy*, **20(4)**, 367–379.

Wright, D.J., Wood, R. and Sylvander, B. (1998). ArcGMT: A suite of tools for conversion between Arc/INFO and Generic Mapping Tools (GMT). *Computers and Geosciences*, **24(8)**, 737–744.

Wright, R.U., Ray, S., Green, D.R. and Wood, M. (1998). Development of GIS of the Moray Firth (Scotland, UK) and its application in environmental management: site selection for an artificial reef. *The Science for the Total Environment*, **223**, 65–76.

Yang, C.R. and Tsai, C.T. (2000). Development of a GIS based flood information system for floodplain modelling and damage calculation. *Journal of the American Water Resources Association*, **36(3)**, 567–577.

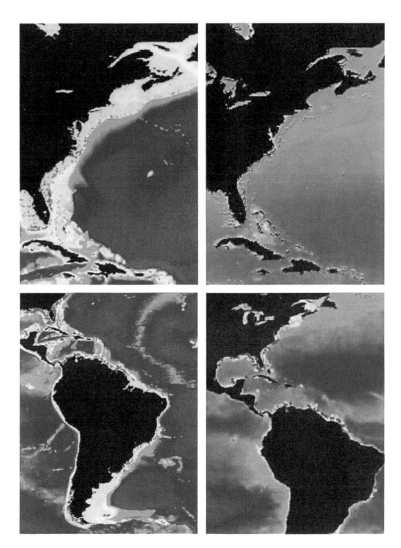

**Colour plate 1.** Ocean Biogeographic Information System (OBIS) is a database of global marine animal and plant distributions. Overlays among data on species geodistribution and environmental conditions are one major output of OBIS. Figures show tonguefish geodistribution (*Symphurus* spp.) combined with bathymetry (left) and primary productivity (right). OBIS is developed at the Institute of Marine and Coastal Sciences at Rutgers University, New Jersey, USA. Figures are courtesy of Phoebe Zhang and Jen Gregg (OBIS homepage: http://marine.rutgers.edu/OBIS/).

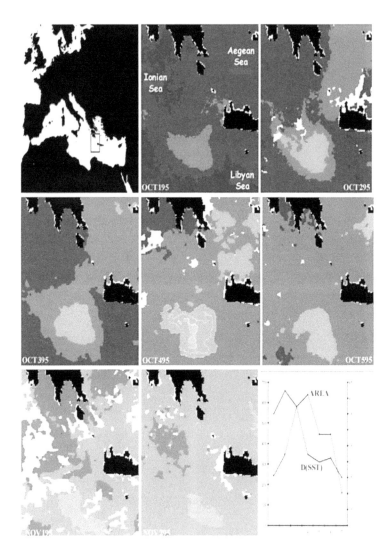

**Colour plate 2.** Examination of gyre formation in SE Mediterranean (West Cretan Gyre) through GIS. Seven sequential georeferenced AVHRR SST images are classified into polygon areas (e.g. OCT495) according to image values (here every 2 °C). Use of polygon classification permits the measurement of gyre spatial extent and difference in SST, D(SST), inside and outside the gyre area (lighter colours show colder SST). The graph shows the observed patterns in D(SST) and area during the 7 subsequent weeks in that fall 1995 gyre event (a notable average of 2.7 °C D(SST) occupying an average area of 4000 m²).

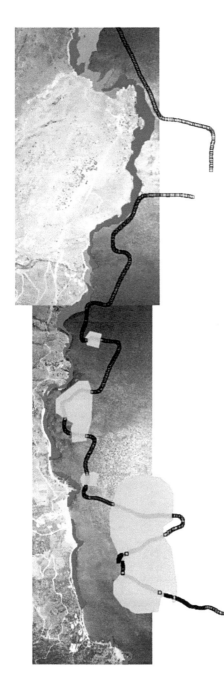

**Colour plate 3.** Example of RoxAnn® sonar sediment data processing and aerial photography interpretation (API) through GIS. This figure is a snapshot from an ongoing process for coastal submerged vegetation GIS mapping of NATURA marine sites in Greece. API includes georeference of aerial photos and their use as basic theme for on-screen polygon digitising of coastal features (vegetated rocks, bold rocks, beaches, etc.). Sonar point data are interpolated to reveal, for example, *Posidonia oceanica* beds (in green) in deeper waters where API is not possible.

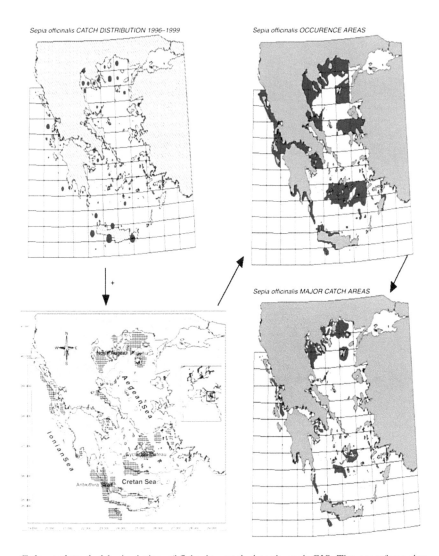

**Colour plate 4.** Manipulation of fisheries catch data through GIS. The georeferencing of fisheries catch data (top left) opens new ways for further data processing. For example, when fisheries catch data is integrated with bathymetry (lower left) can reveal species occurrence areas (top right). Then, occurrence areas can be integrated with major fishing locations (lower left) to reveal species major catch areas (lower right). In such GIS data integrations, species life history data (e.g. maximum depth of species occurrence) is important.

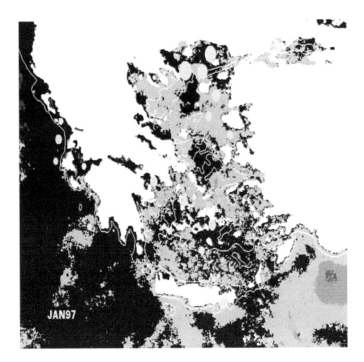

**Colour plate 5.** Overlay among January 1997 cephalopod catch, January 1997 SST anomaly and 600-m bathymetric contour (limit of trawling) in SE Mediterranean. Cephalopod catches are highly associated with the boundaries of SST anomalies (strong indication of fronts or upwellings) and the boundaries of the continental shelf.

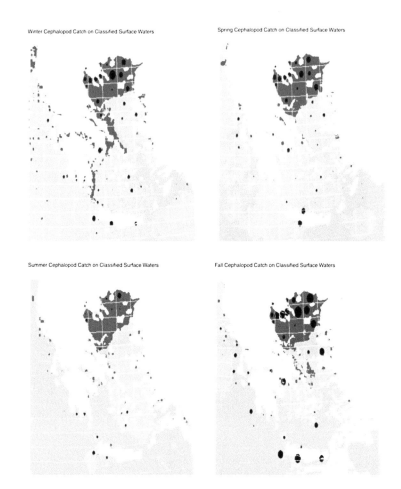

**Colour plate 6.** Total cephalopod catch overlaid on classified surface waters in SE Mediterranean for the period 1997–1999. Classification of surface waters in clusters of similar ranges in temperature, CHL and salinity allows the study of seasonal changes in water masses, which in turn, greatly affect the geodistribution of environmentally sensitive species populations.

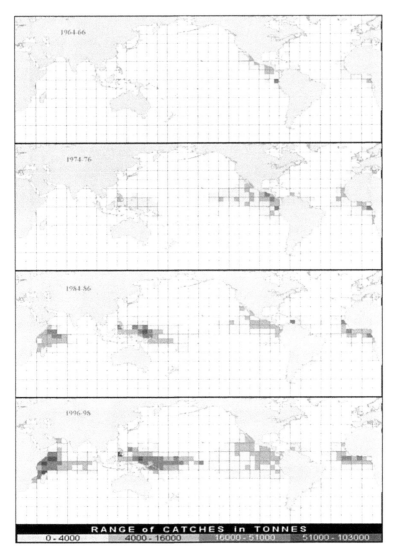

**Colour plate 7.** Atlas of Tuna and Billfish catches developed by the Food and Agricultural Organization of the United Nations (FAO). This online atlas became possible through the cooperation of several nations, which provided FAO with species historical catch data. Tuna and billfish catches by purse seine from 1964 onwards are shown. Images are courtesy of Fabio Carocci, Jacek Majkowski and Francoise Schatto, FAO Fisheries Department, Rome, Italy (Carocci and Majkowski 2001). Atlas homepage: http://www.fao.org/fi/atlas/tunabill/english/home.htm.

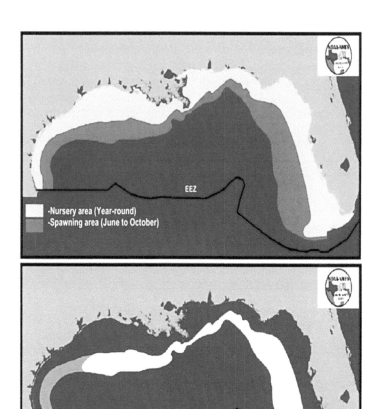

**Colour plate 8.** Essential Fish Habitat (EFH) for Red Snapper (*Lutjanus campechanus*) in the Gulf of Mexico. The US Magnuson/Stevens Fishery Conversation and Management Act of 1996 requires the regional Fishery Management Councils and the Secretary of Commerce to describe and identify EFH for species under federal Fishery Management Plans. EFH is defined as 'those waters and substrate necessary to fish for spawning, breeding, feeding, or growth to maturity'. Images are courtesy of Geoffrey Matthews, US National Marine Fisheries Service, Galveston Laboratory through the Essential Fish Habitat Project: http://galveston.ssp.nmfs.gov/efh/. (See also at: http://christensenmac.nos.noaa.gov/gom-efh/).

Yetter, C. (2000). Development of Delaware's coastal zone environmental indicators decision support system. In *Proceedings of 2000 ESRI International User Conference*. On line: http://www.esri.com/library/userconf/proc00/professional/papers/PAP221/p221.htm

Young, R.S. and Bush, D.M. (2000). On the state of coastal hazards mapping. *Environmental Geosciences*, **7(3)**, 117.

Yang, X., Damen, M.C.J. and Van Zuidam, R.A. (1999). Use of Thematic Mapper imagery with a Geographic Information System for geomorphologic mapping in a large deltaic lowland environment. *International Journal of Remote Sensing*, **20(4)**, 659–681.

Zeidler, R.B. (1997). Climate change vulnerability and response strategies for the coastal zone of Poland. *Climatic Change*, **36(1–2)**, 151–173.

Zheng, Q., Yan, X.H., Liu, W.T., Klemas, V. and Sun, D. (2001). Space shuttle observations of open ocean oil slicks. *Remote Sensing of Environment*, **76(1)**, 49–56.

# CHAPTER THREE

# GIS and Fisheries

## 3.1 INTRODUCTION

The oceans, which cover 71 per cent of the Earth's surface, sustain one of the most precious food resources on our planet. Fish is an important source of food for the sustainability of an increasing human population currently providing the major source of protein for one billion people worldwide. The facts that marine resources are reproducing themselves to provide future generations with food and that the access and harvest of these resources became an easy task resulting to overexploitation makes it important to manage fisheries better. Today, most problems in fisheries are in the spatial domain, since specific areas are heavily fished, habitats are damaged and small size fish are caught. Geographic Information Systems (GIS) as a monitoring and management tool as well as an intermediate process in integrated decision aid tools are exactly designed to solve spatiotemporal problems and are already being used in fisheries applications since the last decade.

During the past decades, the understanding of fisheries has improved considerably. However, this has happened in line with increasing overexploitation of the resource and coupled with other environmental problems, like ocean dumping and pollution has closed some fisheries. Fish, as a food resource, has been gradually endangered as fish harvesting and processing technology has developed rapidly. Inventions such as the power block and the purse seine, fish finding electronics, the extension of trawling to 1000 m of depth, have turned most unknown fish species out of their hiding places. Since the 1970s and in line with these developments, the international community reacted by slowly creating a new fishery administration setting according to fishing craft possibilities. The general extension of the Exclusive Economic Zones (EEZ) to 200 miles from 1977, the adoption of UNCLOS in 1982, the UNCED Agenda 21 in 1992, the UN Conference on Straddling Stocks and Highly Migratory Fish Stocks, the Agreement to Promote Compliance with International Conservation and Management Measure by Fishing Vessels in the High Seas (the Compliance Agreement) in 1993, the 1995 Kyoto Declaration and the FAO Code of Conduct for Responsible Fisheries adopted in 1995 are all testimonies of an increasing concern about our commons and about the way we have been treating our food resources from the oceans.

Although the oceans cover 71 per cent of the Earth's surface and are on average more than 2 miles deep, they are not uniformly populated with fish. In terms of depth, the region of primary production is the near surface layer through which sufficient light can penetrate, called the euphotic zone. In productive coastal waters it is typically 10–30 m deep. Here, solar energy allows the growth of the minute, floating aquatic plants called phytoplankton that serve as a first link in an ensuing food chain. Commercial landings are mostly taken in coastal waters where the greatest density of fish are found in estuaries and embayments that extend outward from the coastline to the edge of the continental shelf. Coastal waters occupy less than 10 per cent of the ocean surface and 1 per cent of its volume but account for nearly a quarter of oceanic biological production, which in turn supplies 90 per cent of the world's fish catch. The rich nutrient content in these waters, seen as enhanced concentrations of chlorophyll in satellite images of ocean colour, is the result of run-off from the land, upwelling, nutrient regeneration and other ocean processes that nourish life in the sea. For example, only about 1/1000 of the oceans' surfaces have natural upwelling, but these areas account for nearly half (44 per cent) of the world's fish food. At the same time, there are great expanses in mid-ocean that yield large quantities of tuna and other highly migratory species. A well-known example is the equatorial eastern Pacific, a common fishing ground for purse seines and longline vessels that catch various species of tuna.

In a setting of extended jurisdiction, nations now enjoy exclusive fishing rights in ocean areas (EEZ) that lie off their coastlines, which are restricted zones expanded by most coastal nations in the mid 1970s to reach 200 miles outward from their shores. These EEZs sustain about 90 per cent of the total world fisheries catch. In principle, the concept of extended jurisdiction solved most of the problems of open access by providing authority to national states to manage the stocks under their jurisdiction, although it is not always used to best advantage. A need to limit access to EEZs is well recognised and a number of schemes have been developed internationally to remedy open access harvesting in EEZs, although such methods have yet to be applied in many of the fisheries around the world. Another relatively recent effort to create spatial marine areas for long-term monitoring and management of marine resources is the concept of Large Marine Ecosystems (LME, http://www.edc.uri.edu/lme/) identified by Sherman *et al.* (1990, 1993). Today, coastal oceans are divided into 50 LMEs and despite that these areas do not provide any official sort of management unit, many research and monitoring initiatives are supported by NOAA (National Oceanic and Atmospheric Administration), IUCN (The World Conservation Union), the University of Rhode Island, ICES (International Council for the Exploitation of the Sea) and IOC (Intergovernmental Oceanographic Commission).

In the last decade, GIS technology, marine RS and fisheries monitoring methods provided a set of data and tools for the management of fishing fleets and fisheries resources in the various EEZs. Fisheries authorities provide time series of fisheries catch data through monthly sampling in various statistical areas that are established in EEZs worldwide. On the other hand, many marine species population characteristics were thoroughly studied through biological and genetic research. In a continuing effort, fishery biologists discover and put together the most important pieces of the puzzle of marine species life cycles. With help from geneticists, who categorise different species of the same genus that share common biotopes, define the spatial extent of occurrence of certain species. In addition, the monitoring of fishing fleet activity has provided valuable information on the effectiveness of various fishing tools as well as a picture of the major fishing grounds.

All this information on species populations is listed as a species life history data. These data provide information on species type (e.g. benthic or pelagic), species preferred living ranges of temperature and salinity, recruitment periods, spawning periods and characteristics (e.g. preferred spawning sediment types and spawning temperature and depth ranges), migration habits, maximum depth of species occurrence, etc. For example, life history data on four cephalopod species are shown in Table 3.1. In studies of species population spatiotemporal dynamics, species life history data may be viewed as a starting point for GIS analysis, particularly when combined with satellite RS. GIS use this information as constraint parameters in the analysis of environmental and fisheries data and provides integrated output on seasonal areas that are important in various stages of species life cycles. GIS may reveal the geographic distribution of species life history data and in combination with results from oceanographic GIS analysis, may reveal the dynamic interactions between species populations and oceanographic features in a spatiotemporal scale. Specific spatial questions on species resources dynamics, such as: where do they spawn, what are their migration corridors, where are the areas of their occurrence, where do they mainly fished, what is the geodistribution of their abundance, where are their seasonal suitable habitats, questions including also the temporal context, may be examined with the use of GIS. Indeed, GIS technology contributes in the complicated fisheries management process by processing fisheries monitoring data to meaningful information (e.g. species geodistribution maps), integrating monitoring and environmental data, enhancing fisheries monitoring procedures and providing integrated output to fisheries managers.

**Table 3.1.** Life history data on the ecology and biology of four cephalopod species organised and provided by the International Council for the Exploitation of the Sea (ICES).

| SPECIES<br><br>LIFE HISTORY | LONG-FINNED SQUID (*Loligo vulgaris*) | CUTTLEFISH (*Sepia officinalis*) | COMMON OCTOPUS (*Octopus vulgaris*) | SHORT-FINNED SQUID (*Illex coindetii*) |
|---|---|---|---|---|
| Benthic/Pelagic | PELAGIC | BENTHIC | BENTHIC | PELAGIC |
| Temp. range | 10–25 °C | 10–30 °C | 10–30 °C | 7.5–20 °C |
| Spawn. season | DEC–JAN | MAR–JUL | JUN–SEP | SPRING/AUTUMN |
| Spawn. depth | 10–30 m | 2–50 m | 100 m | UNKNOWN |
| Substrate type | HARD | MUDS/SANDS | ROCKS/SANDS | UNKNOWN |
| Bathymetry range | 10–100 m | 10–300 m | 0–500 m | 60–250 m |
| Migration pattern | IN/OFFSHORE | OFF/INSHORE | OFF/INSHORE | UNKNOWN |
| Migration scale | 200 Km | 50 Km | 50 Km | UNKNOWN |

As in the case of oceanographic GIS, RS data and the Internet play important roles in fisheries GIS development. In fact, EO data is the main factor that merges oceanographic and fisheries GIS applications in an effort to better understand the dynamic interactions between species populations and the marine environment. Soliday (2000) discussed the various new and unexpected uses of high-resolution satellite imagery, especially for the protection of marine habitats as these are related to fisheries. Satellite imagery is extensively used in fisheries GIS studies for the identification of relations among catch data and environmental variables, such as SST distribution and CHL concentration. In addition, RS data is used in several studies for the mapping of species essential habitats including spawning grounds. Simpson (1994) and an excellent online report by Scientific Fishery Systems, Inc. (1999) illustrate the fisheries applications of satellite imagery while Mumby *et al.* (1999) studied the cost effectiveness of RS imagery for tropical coastal resources assessment and management. In general, wide use of AVHRR SST is made in both commercial and recreational fisheries because certain species have been found to aggregate near thermal ocean gradients (fronts, upwelling). Temperature has had a tremendous impact on the efficiency of US tuna fleets in the Pacific, often reducing search times by 25–40 per cent (Simpson 1994). Surface CHL concentration (CZCS and SeaWiFS) is also widely used, especially in combination with temperature. Areas of relatively stable temperature and high CHL concentrations may attract feeding pelagic species. MODIS is another sensor of which acquired data are used in fisheries. MODIS combines the functions of AVHRR and SeaWIFS and greatly facilitates combined view of surface ocean conditions.

The Internet, on the other hand, plays an important role on the dissemination of GIS data and output and provides a platform for extensive fisheries services to the public. In cooperation with NASA, the Colorado Centre for Astrodynamics Research provides satellite altimetry to Scientific Fishery Systems, Inc., who combines this data with other GIS data to provide fishermen with bathymetry, water temperature and historical catch statistics through the Internet and stand-alone application software. These products also allow fishermen to log their catch, tides, currents, marks, ocean temperature, weather conditions and lunar state. In addition, among many similar companies, Roffer's Ocean Fishing Forecasting Service, Inc. (ROFFS™, http://www.roffs.com/), a scientific consulting company based in Miami (Florida) provides worldwide fisheries oceanographic consulting services. They combine fisheries data with satellite and other oceanographic data to produce tactical and strategic fisheries forecasts. Their principal product is the 'ROFFS Oceanographic Fishing Analysis', which is designed to allow fishing vessels to concentrate fishing effort in the most productive waters. The analysis incorporates numerous factors including: water temperature, water colour, orientation of local currents, history of ocean fronts, bottom topography, biological quality of the water, forage preference of the target species, availability of forage and habitat preference of the forage and target species.

Another discipline commonly applied in fisheries GIS applications is that of Geostatistics. Geostatistics are statistical methods used to describe spatial relationships among sample data, predict spatiotemporal phenomena and interpolate values at unsampled locations. Geostatistics have traditionally been used in geosciences, such as meteorology, mining, soil, forestry, fisheries, remote sensing and cartography. Geostatistical techniques (objective analysis) were originally developed by Soviet scientists for meteorological data predictions (Gandin 1963). In Fisheries, these techniques are mainly used for the estimation of fish stocks and the relations between the spatial distribution of fish species and environmental variables (e.g. Petitgas 1993; Williamson and Traynor 1996; Fletcher and Sumner 1999; Bez and Rivoirard 2000; Georgakarakos *et al.* 2002). Rivoirard *et al.* (2000) mentioned that geostatistic techniques are widely recognised as an important tool for the estimation of the abundance and distribution of natural resources with new geostatistical developments applied to fisheries science, particularly in variogram estimation. 'Practical Geostatistics', published by Isobel Clark in 1979 may be freely downloaded by Ecosse Geostatistical Sales' website (http://geoecosse.hypermart.net/), which also offers a new 2000 update of the same book.

## 3.2 WORLDWIDE FISHERIES GIS TOOLS AND INITIATIVES

Currently, GIS is used in a variety of ways in fisheries (inland and marine) and aquaculture fields and during the last 3 or 4 years has been adopted as database and visualisation methods by many worldwide organisations. The Fisheries Global Information System (FIGIS) is an Internet-based interactive system on worldwide fisheries aiming to provide policy makers with timely, reliable strategic information on fishery status and trends on a global scale. FIGIS is a project developed by the Food and Agricultural Organisation of the United Nations (FAO), Fisheries Data and Information unit (FIDI) for the Fisheries Division. FIGIS, which at this time is under development, will provide spatial and other information on aquatic species (e.g. life history data), marine resources and marine fisheries (e.g. data, major issues, multimedia documents) and fishing technologies (gears, vessels, fishing techniques, fleet statistics) through FAO website (http://www.fao.org/fi/figis/) in three different languages: English, French and Spanish. FIGIS will support policy makers' shifting towards sustainability centred management by providing them with a single entry point to strategic data, information, analyses and reviews of fisheries issues and trends. FIGIS will allow the user to access a variety of domains including biology, fishing technology, high seas vessel records, resources, fisheries, management systems, aquaculture, products and markets. Regional Fishery Bodies and National Centres of Excellence will update FIGIS databases through agreed standards for vocabularies and classifications for indexing, glossaries to ensure definitions of terms and norms for datasets. FIGIS will put emphasis on the geographical aspects of data by relying on a powerful GIS.

FAO also established an online worldwide Atlas of Tuna and Billfish Catches with excellent GIS maps of fisheries information about these species (Carocci and Majkowski 1996 and 2001). The preparation of this atlas became possible through the collaboration of numerous institutions and scientists from Australia, Seychelles, Spain, New Zealand, USA and New Caledonia. The atlas provides worldwide average annual catches of tuna and billfish (Figure 3.1, see also Colour plate 7) for three fishing gears (longline, purse seine and poleline) and may be accessed online (http://www.fao.org/fi/atlas/tunabill/english/home.htm).

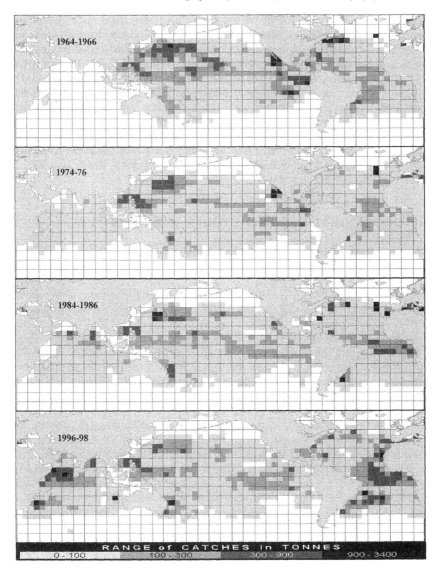

**Figure 3.1.** FAO's Atlas of Tuna and Billfish Catches. Longline billfish catches from 1964 onwards are shown. Images are courtesy of Fabio Carocci, Jacek Majkowski and Francoise Schatto, FAO Fisheries Department, Rome, Italy (Carocci and Majkowski 2001).

NOAA's National Ocean Service initiated a number of projects under the Biogeography Programme (http://biogeo.nos.noaa.gov/) aiming to develop knowledge and products on living marine resource distributions and ecology throughout the US estuarine, coastal and marine environments. Activities of this important programme focus on developing products, applications and processes for defining and interpreting the relationships of species distributions and their environments. The major topics of this programme include benthic habitat mapping in coral reef areas, aerial photography for the US Caribbean and Hawaii, habitat suitability modelling using GIS, distribution of abundance and life history of estuarine fishes and invertebrates, coastal ocean strategic assessment data atlases, and development of custom GIS applications for resource assessment and modelling supporting NOAA's Essential Fish Habitat mandate (EFH, http://www.nmfs.noaa.gov/ess_fish_habitat.htm).

The mapping and conservation of EFH is a recent thrust in US national fisheries management (Figure 3.2, see also Colour plate 8). Under the 1996 Sustainable Fisheries Act, an amendment to the Magnuson/Stevens Fishery Conservation and Management Act (Public Law 94/265, http://www.nmfs.noaa.gov/ sfa/magact/), US Congress stipulated provisions to include identification and management of fish habitat. Specifically, the Sustainable Fisheries Act defined EFH as 'those waters and substrate necessary for the spawning, breeding, feeding or growth to maturity' of fish and invertebrate species, establishing guidelines to assist regional fishery management councils to describe and identify EFH in their fishery management plans. Such plans would minimise the harm to EFHs caused by fishing and identify other actions to encourage the conservation and enhancement of EFHs. In this effort, GIS is widely used to visually overlay species distribution with habitat features providing a powerful tool for classifying EFHs including habitat quality and habitat suitability indices, optimum sampling surveys and population abundance estimates.

The Fisheries Centre at the University of British Columbia (FC, http://www.fisheries.ubc.ca/) provides an extensive set of products on fisheries, including online publications (books, journals, FishBytes and project reports) and software, such as the larval drift and growth simulator (Walters *et al.* 1992). Ecopath with Ecosim (http://www.ecopath.org/) is another fisheries related set of software, which is designed to help construct models of trophic flows and interactions in marine and freshwater ecosystems. The Ecopath activity is an ongoing international effort between the National Marine Fisheries Service (NMFS, Hawaii) and the International Centre for Living Aquatic Resources Management (Manila, Philippines) supported by the Danish International Development Assistance (Danida). There is an extensive list of published Ecopath models for many areas of the world's oceans and inland waters. Another software developed by FC is Ecoval, a new decision-support technique that allows fisheries policy makers to compare the ecological, economic and social benefits of different states of a marine ecosystem and establish a basis for monitoring management goals. Ecoval is closely used with Ecopath and Ecosim for a comprehensive understanding of marine ecosystems from both ecological and socio-economic point of views.

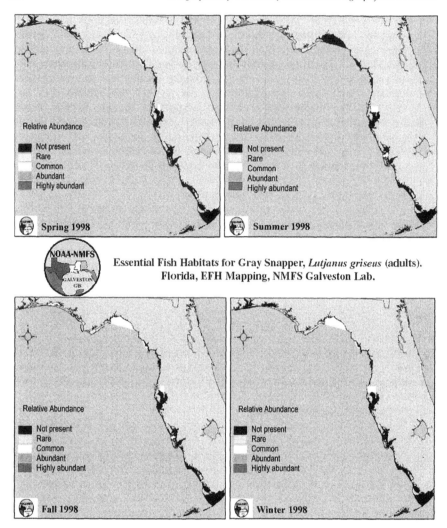

**Figure 3.2.** GIS output on seasonal EFH for gray snapper in Florida, USA. The US 1996
Magnuson/Stevens Fishery Conservation and Management Act established guidelines to assist regional
fishery management councils to describe and identify EFH in their fishery management plans. Images
are courtesy of Geoffrey Matthews, US National Marine Fisheries Service, Galveston Laboratory
through the EFH Project: http://galveston.ssp.nmfs.gov/efh/ (See also at:
http://christensenmac.nos.noaa.gov/gom-efh/).

The Fisheries Information and Analysis System (FIAS) will be a product of the ongoing project 'Ecosystem based approaches to resource management in Northwest Africa' funded by the European Commission. Participants include FAO, four research institutions based in Europe (Italy, France, Spain, Portugal) and six research institutions in NW Africa from the SubRegional Fishery Commission (Guinea, Guinea/Bissau, Senegal, the Gambia, Cape Verde and Mauritania). The FIAS project (http://193.43.36.44/fi/projects/FIAS.asp) aims in strengthening the capacities for improved resource management in the partner countries in NW Africa, both at national and regional level. European partners contribute to the project with the development of scientific tools (database systems) and the repatriation of information held in Europe (surveys, catch and effort data, etc.) while African partners are systematically identifying available information within their reach (publications, archives) and make it available to the project. FIAS is implemented through a number of interlinked modules including a GIS where fisheries related data and other datasets (survey data, geographic maps, oceanographic data, RS data) will be brought together for presentation and for an improved understanding of spatial processes affecting fisheries. This GIS module will be linked with other modules on the biology and ecology of demersal species (FISHBASE), evaluation of resource trends from surveys (TRAWLBASE) and ecosystem modelling (Ecopath and Ecosim).

The VIBES Project (Viability of exploited pelagic fish resources in the Benguela Ecosystems in relation to the environment and spatial aspects) is an applied 4-year research programme (1997–2001) directed towards providing new tools and information for regional assessment of coastal pelagic resources and their management in the Benguela ecosystem. The main objective of VIBES is to study long-term changes in fisheries, pelagic fish recruitment and their spatial patterns in relation to the environment, and adult fish abundance according to different scales of observation. The project's has two main scientific research tasks: (1) fish population spatial dynamics and recruitment variability; and (2) environmental processes. VIBES products include a GIS as an aid for assessment of the relationships between the environment, the resources and exploitation, methods and software for quantifying an upwelling index using RS images, and a preliminary modelling of biological and environmental processes in the area of St Helena Bay (Freon *et al.* 1999). VIBES is a collaborative project between IRD (Institut de Recherche pour le Developpement), SFRI (Sea Fisheries Research Institute), UCT (University of Cape-Town) and other universities and institutes in the region of South Africa.

The International Research Institute for Climate Prediction (Columbia Earth Institute, Lamont/Doherty Earth Observatory, Palisades, NY) use satellite information from the Advanced Very High Resolution Radiometer (AVHRR) sensor, the European Remote Sensing Satellite (ERS-2), the NASA Scatterometer (NSCAT), the Ocean Colour and Temperature Sensor (OCTS), and the Sea viewing Wide Field of view Sensor (SeaWiFS) to study the evolution of oceanographic conditions off the coast of Peru comparing observations with the Peruvian catch of small pelagic fish. The potential of these satellite observations linked to a planktonic ecosystem model for the quantification of the impact of El Nino conditions on pelagic fish catch as well as on the societal utilisation and impact of such information are the main goals of this monitoring effort. A result of this effort was the temporary closing of fishery in July 1997 to protect the remaining stock based on the estimated correlation among fish catch and sea-level anomaly, SST, the SST difference (which indicates upwelling intensity) and satellite chlorophyll, as well as on two output variables from an ecosystem model, phytoplankton biomass and carbon available for fish.

The Great Lake Fisheries GIS, a continuing monitoring and analysis project initiated by the Institute for Fisheries Research, School of Natural Resources and Environment, University of Michigan, is a well-established monitoring and management system for fisheries resources of the Great Lakes. The area is divided in a $10 \times 10$ inches statistical 'rectangle' monitoring system linked to an extensive GIS database. Output analysis maps include the geodistribution of harvest, effort and tagging data for a great number of species, modelling of species dense capture areas, calculation of distances between stock and capture sites, chinook salmon, walleye, lake trout and steelhead tagging and recapture sites, effects of rivers and dams on species distributions, landscape summaries of precipitation, bedrock and surficial geology, classification of aquatic habitats and development of spatial models of growth and movement for salmonines, yellow perch and walleye (Rutherford and Brines 1999).

In a European context, the Commission of the European Communities' Common Fisheries Policy, although being organisation and structure oriented, clearly introduces a number of conservation actions in order to protect fisheries resources by regulating the amount of fish taken from the sea and by allowing young fish to reproduce (CFP, http://europa.eu.int/comm/fisheries/doc_et_publ/ cfp_en.htm). CFP, which is currently under review, calls for integrating environmental considerations (e.g. habitats) and factors affecting fishing (e.g. pollution) in fisheries policy-making. Use of GIS for such calls is profound. Caiaffa (2000) investigated the potential of GIS as a tool for the assessment of the European seas and presented EUMARIS GIS (funded by the European Environment Agency), which includes many examples of thematic maps for different marine subjects. Also, MarLIN (Marine Life Information Network, http://www.marlin.ac.uk) is an initiative of the Marine Biological Association of the UK (http://www.mba.ac.uk/) and in collaboration with major national and international holders and users of marine biological data provides the most comprehensive and easily used source of information about marine habitats, communities and species. MarLIN describes approaches for the identification of 'sensitivity' and 'recoverability' in species and biotopes through current development of GIS databases. The Fisheries GIS Unit (Canterbury Christ Church University College, UK) is a highly active unit involved in fisheries GIS developments worldwide (e.g. French Guiana, Sri Lanka, Vietnam). They developed FISHCAM, an electronic fisheries data log system based on GIS, which is used in many fisheries around the world (http://www.cant.ac.uk/depts/acad/ geography/fish/fishcam.htm).

The FAO COPEMED Project (1996–2001) among Algeria, France, Italy, Libya, Malta, Morocco, Spain and Tunisia, aiming at the enhancement of scientific knowledge for a better management of Mediterranean Fisheries, excels the use of GIS in fisheries management and provides invaluable information on how a spatial approach using GIS may help decision makers. Taconet and Bensch (1999) summarised the COPEMED GIS approach underlining the importance of the spatial dimension in fisheries management and of the location of resources under exploitation, the need for spatial modelling and the need of linking GIS with other database management systems, geostatistics, multidimensional statistics packages and numerical models. They also mention the use of GIS for the identification of highly localised fisheries related problems, since the different socio-economic components and the distribution of exploitable resources greatly vary from one area to another. Since the main purpose of management measures is to prevent the resources from being overexploited, mapping geomanagement areas, such as resource distribution, localised accessible areas, authorised areas (according to regulations in force) and fleet activity extent is a useful tool in order to simulate the effects of prospective management measures.

## 3.3 GIS APPLICATIONS IN MARINE AND INLAND FISHERIES AND AQUACULTURE

Knowledge on the spatiotemporal distribution of fish populations and their habitats as well as on how fish populations interact on seasonal changes in the marine environment is important for modelling their population dynamics and managing fishery resources. GIS tools process spatial information from many sources in order to examine distribution patterns on a wide scale, allow researchers to detect changes in distribution, examine connectivity among species and life stages and investigate determinants of geographic distribution. Developments of fisheries GIS applications are rapid and include the identification of discrete geographic stocks, description of EFH, improvement of survey designs and delineation of optimal areas for fishing, aquaculture and marine reserves. Isaak and Hubert (1997) outlined the advantages and costs of developing GIS for fisheries applications and described four categories of applications depicting the questions: (1) what are the attributes of a location; (2) what areas meet a specific set of criteria; (3) what are the spatial patterns; and (4) what happens if a management action is implemented. They concluded that most fisheries GIS developments are limited to the first type of application (developments 'back' in 1997), although today most fisheries GIS are up to the third type of applications, progressively proceeding towards the generation and evaluation of different management schemes. Nowadays, more advanced features of GIS are employed and emphasis shifts from descriptive uses to exploratory applications.

Currently, fisheries GIS literature on peer-reviewed journals is relatively limited, although grey literature is notable (Meaden 2000). An important reason that there are not many dynamic GIS applications in the primary literature is that journal publications accept geographic charts that can effectively illustrate an observed pattern, however publishing GIS results on dynamic patterns or for evaluation of different scenarios is more difficult to express in such a form. Xavier *et al.* (1999) referred the journal reader to a website where readers can access results by querying the various databases and accessing hundreds of different views. This approach is growing and becoming an effective way to publish dynamic GIS results. With electronic publications becoming more common day by day, maybe it would not be too late when dynamic fisheries GIS applications will appear as parts of electronic publishing. Although many fisheries GIS-related workshops have taken place as satellite meetings in major marine conferences, only three worldwide symposia on Fisheries GIS have already organised with great scientific response. In 1999, the First International Symposium on GIS in Fishery Science (2–4 March 1999, Seattle USA, http://www.esl.co.jp/Sympo/sympo.htm) gathered some 230 participants from 28 countries, 90 oral presentations, 41 posters and 10 software demonstrations. The symposium's proceedings include summaries of all presented material as well as 39 selected full papers, providing readers with ideas for potential GIS-related work (Nishida *et al.* 2001). The conference included application examples for a wide range of geographic areas, from relatively small areas (e.g. streams, watersheds, lakes) to larger areas (e.g. Gulf of Alaska, Mediterranean Sea, New Zealand and South African waters, Georges Bank, Japan Sea). Presentations also covered a broad range of GIS methodologies including simple mapping, multidimensional presentations, sophisticated overlay techniques, spatial numerical analyses and RS data integration for many aquatic and marine species and life stages (e.g. catfish, shellfish, kelp, small pelagic fish, epibenthos, demersal fish, highly migratory species). Later in 1999, the 17th Lowell Wakefield Fisheries Symposium on Spatial Processes and Management of Fish Populations (27–30 October 1999, Anchorage Alaska, http://www.uaf.edu/ seagrant/Conferences/Spatial-reg.html) gathered some 110 presentations in eight parallel sessions, which included presentations on analytical tools, various modelling approaches, patterns in species life history and spatial distribution of populations, relations with the environment, interactions among species and design of marine protected areas (MPAs). Abstract proceedings are freely available by Alaska Sea Grant College Programme, University of Alaska Fairbanks. In 2000, the Fisheries GIS Symposium for the American Fisheries Society (20–24 August 2000, St. Louis USA, http://www.ecu.edu/org/afs/st_louis/GISsymposium.htm) gathered 18 oral presentation, which are available online. Presentations included fisheries GIS applications for stream, river, lake and marine environments. Topics were related to integrating hydrology and ecology for the development of GIS models, identifying targets for the conservation of aquatic biodiversity and predicting stream habitats and fish diversity. The Second International Symposium on GIS/Spatial Analyses in Fishery and Aquatic Sciences (to be held at the University of Sussex, Brighton, United Kingdom, 3–6 September 2002, http://www.esl.co.jp/Sympo/sympo10.htm) will probably gather recent GIS developments in the field, since during the last 3 years, there is a tremendous technological and scientific development in oceanographic and fisheries GIS.

As an output from the First International Symposium on GIS in Fishery Science, Meaden (2000) listed the main thematic areas into which GIS will permeate. These areas include the management of catch and effort data, the establishment and monitoring of EFH and marine reserves, the management and monitoring of stock enhancement programmes, the development of trawl and fisheries impact assessments, the mariculture and aquaculture location and impact studies and the monitoring and modelling of fish stocks and fishery quotas. All these thematic areas have a strong spatial background and the current use of GIS, as the natural framework for spatial data management, will definitely expand in the future. The time has arrived for a Fisheries GIS journal and a part of the ideal organisation of such a publication would be to serve multiple output or dynamic fisheries GIS applications through a specifically dedicated Internet server.

During the last decade, marine scientists, fisheries managers and the general public are becoming increasingly concerned about the direct, indirect and cumulative impacts of habitat change on commercial and recreational fisheries and the effects of these fisheries from an ecosystem perspective. These concerns are coupled with declines in some fish stocks due to mismanagement, such as overfishing, failure to account for bycatch, gear damage to habitats and changing environmental conditions. In parallel, there has been an enormous scientific effort to examine the spatial dynamics of marine populations in relation to those of marine environment, identify species important habitat areas and disseminate the importance of such dynamic relations to official management authorities. Authorities, in turn, have started to realise the importance of the spatial component in marine dynamics (e.g. US Sustainable Fisheries Act of 1996, European Common Fisheries Policy) and in many cases, to include this spatial component in fisheries management policies (e.g. EFHs, MPAs). Under this setting, GIS offers the appropriate technology for the generation of tools to support spatial management decisions in the fisheries sector and it is increasingly being used in many fisheries worldwide.

### 3.3.1 Marine Fisheries

In marine fisheries studies, GIS is developed in a variety of ways, including all thematic areas mentioned above and identified by Meaden (2000). Kieser *et al.* (1995) enumerated the advantages and disadvantages of using GIS in fisheries science and demonstrated the strength of using GIS visualisation in studies of fisheries acoustic scatter data, especially when these data are combined with various environmental parameters. Stolyarenko (1995) proposed that GIS can be used to optimise survey stratification and furthermore that on-board GIS systems can be used to adaptively modify sampling designs as data are collected. The proposal involved loading data to a spatial model (spatial regression or spline techniques) to indicate optimal sampling strata. The method aimed to facilitate the common fact that survey observations are made by discontinuous sampling at randomly selected locations in a marine system. Caddy *et al.* (1995) examined 1970–1989 trends in fish landings in the Mediterranean and Black Sea basins for each statistical area of the General Fisheries Council for the Mediterranean (GFCM) by comparing estimates of production in different basins using the productivity in tones per surface area of continental shelf. Landing trends were related to human-induced environmental trends (increase in nutrient levels) for areas such as the Black and Adriatic seas (nutrient enrichment from land and river run-off), the Aegean Sea (nutrient rich water inflow from the Black Sea) and the Gulf of Lions (inflow from the Rhone river).

In 1995, technical cooperation between Japan's International Cooperation Agency and Bahrain's Directorate of Fisheries of the Ministry Works and Agriculture resulted in the design of a GIS-based development called Shrimp Fisheries Information System. As fishing stocks have been observed to be diminishing, the Government of Bahrain has been studying a variety of measures to monitor these fishing stocks and place the necessary controls as well as give the best advice on the seasons for fishing for every species, the best areas for fishing and the types of fishing techniques. GIS provided a tool for monitoring and analysis of production in different fishing areas so that steps can be taken towards the protection of breeding stocks (Mehic *et al.* 1996). Meaden (1996a,b) outlined LIBFISH, a GIS initiative for the management of Libya's marine fisheries concentrating on marine water conditions and habitat, natural marine resources, fisheries management and regulation, fishing effort and catch, marine resource marketing, mariculture and the coastal environment. Ford and Bonnell (1997) compared results from two GIS studies involving extensive mapping of marine fisheries and biological resources to evaluate the various data needs of marine fisheries managers. Service and Magorrian (1997) utilised side scan sonar and underwater video data coupled with GIS techniques to detect temporal and spatial effects of trawl fishery on an epibenthic community associated with horse mussel in a Northern Ireland sea lough. Giske *et al.* (1998) reviewed the available mechanistic modelling tools for the description of spatial distributions of fish populations, especially the modelling of how fish sense and respond to their surroundings based on their vision, olfaction, hearing, their lateral line and other sensory organs. They categorised such models based on their optimisation and adaptation approach. Among optimisation tools, optimal foraging theory, life history theory, ideal free distribution, game theory and stochastic dynamic programming are presented while among adaptation tools genetic algorithms and the combination with artificial neural networks are described. Foucher *et al.* (1998) developed a GIS for the octopus and finfish fisheries on the Senegalese continental shelf using commercial fishing and oceanographic survey data and geographic objects describing the physical and juridical environment, trawl operations and artisanal fishing sites. They identified areas of conflict between artisanal and industrial fisheries and provided alternative explanations for fisheries management on the degree of respect for the limits of regulated fishing areas and spatial fishing unit strategies according to the main fishing seasons. Rubec *et al.* (1998a,b) discussed the intention of the State of Florida to create a database serving to identify and spatially delineate fish habitats. The Florida Estuarine Living Marine Resources System consists of a relational database that summarises bibliographic information on the habitat requirements of fish and invertebrates, which are highly important to fisheries. They implemented the use of Habitat Suitability Index (HSI) in conjunction with GIS technology to establish suitability indices of relative fish abundance across environmental gradients and creating predictive fish and invertebrate distribution maps. Garibaldi and Caddy (1998) used GIS to quantify species richness and identify different faunal regions in an attempt to biogeographically characterise the Mediterranean and Black Seas faunal provinces. They used a binary matrix of 536 species diversity (present or absent) at 1219 systematic locations (a half minute grid of latitude and longitude) to identify a general decrease in number of species from west to east and nine distinct regions with similar communities. Booth (1998) used GIS to locate optimum areas for finfish harvest off South Africa. Survey catch data, bathymetry, bottom temperature and dissolved oxygen were integrated into GIS to investigate species habitats. GIS was used to locate areas with greater than 80 per cent adults for defining optimum harvest areas, with the objective of avoiding harvest of small and immature fish.

In 1999, NOAA's National Marine Fisheries Service (NMFS) developed a GIS for near real-time use of RS data in fisheries management in the Gulf of Mexico (Leming *et al.* 1999). They developed an ArcView-based GIS, which is accessed through the Internet via password protected user accounts. The system integrates AVHRR/SST and SeaWiFS/CHL imagery and NMFS shrimp statistical data in the jurisdictional boundaries between US and state waters and US and international waters. Stanbury and Starr (1999) discussed the applications of GIS to habitat assessment and marine resource management providing a thorough analysis and explanation of the potential benefits of GIS modelling directly applicable to fisheries science and management. They developed a GIS application for the Monterey Bay National Marine Sanctuary in California. The system integrates a variety of marine and terrestrial data and focuses on emergency response to oil spills, habitat monitoring and fisheries depletion. Xavier *et al.* (1999) developed a GIS to describe the geographic distribution of 21 species of squid in the Antarctic combining location and catches from 2497 research survey stations spanning over 100 years (dating back to the 1886 Challenger expedition), bathymetry from a digital atlas, oceanic fronts from a 1995 study and sea ice extent from a 1992 study. Results showed that most species were distributed south of the Antarctic Polar Front and that morphological features affected distribution (e.g. species with similar body features had similar geographic patterns). Output distribution maps are placed on the web (http://www.nercbas.ac.uk/public/mlsd/squid-atlas/). Kracker (1999) discussed several methods and technologies, such as landscape ecology, GIS and RS, on the basis of their use in the investigation of marine and large lake ecosystems. The fields of spatial analysis and fisheries can be bridged by the analysis of the spatial pattern of species distribution within the water column and the impact of that organisation on ecological processes. In 1999, NMFS mapped the cod distribution in the Gulf of Maine illustrating that cod were concentrated in Massachusetts Bay but were scarce in other areas of the Gulf of Maine. Results are presented through the Internet (http://www.nefsc.nmfs.gov/cod99/) in an effort to avoid overfishing of a geographically concentrated resources, a similar pattern preceded the collapse of northern cod fishery. Begg *et al.* (1999) used descriptive plots of fish distribution to show distinct geographic groups and support statistical analyses describing differences in growth and maturity among fishing grounds off New England (Gulf of Maine, Georges Bank and southern New England) providing insight in the stock structure of New England groundfish. Caddy and Carocci (1999) compared the classic geographical 'friction of distance' approach to generating fields of action around fishery ports of inshore fleets making day trips to their adjacent fishing grounds with a 'Gaussian Effort Allocation' modelling approach where peak effort may occur at different distances from port. This GIS-based approach allows a range of geographical characteristics to be analysed for the description of the interactions of ports and local fleets with inshore resources and local fishing grounds and the identification of resource depletion close to ports. Drury (1999) discussed the classification of habitats in four harbours and estuaries near the city of Auckland (NZ) aiming to design future resurveys of the area to assess fish habitats in these areas. Fairweather *et al.* (1999) developed a fisheries information system to aid in the management of fisheries along the west coast of South Africa. The system integrates multiyear research and commercial trawl data for mapping the seasonal abundance of 17 species.

Ault *et al.* (1999) developed a generalised spatial dynamic age structured multistock production model by linking bioenergetic principles of physiology, population ecology and community trophodynamics to a 2D finite element hydrodynamic circulation model. The system identifies animal movements, which are based on the selection of favourable environmental/habitat features. Ward *et al.* (1999) interpreted aerial photographs, existing habitat maps and local knowledge to create species assemblage indices from comprehensive survey data on 977 species taxa and to produce classification maps of plant, fish and invertebrate assemblages. Their aim was to select and design marine reserves in Perth (Western Australia). Price *et al.* (2000) used existing environmental and bioeconomic datasets for the assessment of Cameroon's coastal and marine environment in the area likely to be influenced by the Chad/Cameroon pipeline. This integrated marine assessment included fisheries, a coastal resource of direct significance in Cameroon. De Leiva Moreno *et al.* (2000) created various indices for 14 semi-enclosed marine statistical areas around Europe using GIS techniques. Indices included the ratio of catchment area to sea area, CHL concentration from RS imagery and the degree of geographical enclosure. These indices were compared with the ratio of pelagic to demersal and invertebrate landings (P/D ratio). The study showed that the P/D ratios were ranged from less than 1.0 for nutrient limited seas (e.g. the Aegean and Ionian Seas) to more than 10 for more eutrophic water bodies such as the Black and Azov seas. Rubec *et al.* (2000) used GIS to produce four zones of abundance-based suitability index curves across environmental and habitat gradients (temperature, salinity, depth and bottom type) for Tampa Bay and Charlotte Harbor fisheries. They overlaid catch data for four species to show the relations between catches and habitat zones and seasonal abundance transfer between the estuaries. Booth (2000) presented a spatially referenced spawner biomass per recruit model for the sparid fish *Pterogymnus laniarius* on the Agulhas Bank (South Africa) to illustrate the applicability of incorporating spatially referenced information in providing management advice. Brown *et al.* (2000) mapped habitat quality for eight fish and invertebrate species in Casco Bay and Sheepscot Bay (Maine) in an effort to support a wide range of management needs involving species and habitat mapping, including analysis of EFH. They used GIS to create raster HSI for each bay using published information and expert reviews for two to four life stages of eight marine species. McConnaughey and Smith (2000) mapped spatially explicit relationships between flatfish abundance and surficial sediments in the eastern Bering Sea by integrating species abundance and sediment data as well as flatfish food habits because sediment properties affect the distribution and abundance of benthic prey.

Pauly *et al.* (2000) proposed that a classification system developed by Platt and Sathyendranath (1988 and 1999) and implemented by Longhurst (1998), defined largely by physical parameters and which subdivides the oceans into four 'biomes' and 57 'biogeochemical provinces', could be merged with the system of 50 Large Marine Ecosystems (LMEs), which would represent subunits of the provinces. Combined mapping in these subunits will allow the computation of GIS derived properties, such as temperature and primary production and their analysis in relation to fishery catch data for any study area as well as it will enable scientists to quantify the EEZs of various countries in terms of the distribution of marine features (e.g. primary production, coral reef areas) so far not straightforwardly associated with different coastal states. Valavanis *et al.* (2001) used GIS to integrate catch data, major fishing areas, SST distribution, CHL concentration and salinity to map the spatiotemporal extend of cephalopods' suitable habitats in SE Mediterranean. Life history data for five commercially important cephalopod species were used as constraints parameters in GIS modelling of species essential habitats. Pierce *et al.* (2002) reviewed the latest GIS developments in cephalopod fisheries giving examples of related applications from Europe, US, South Africa, Thailand and SW Atlantic. Pertierra *et al.* (2001) used buffering and kriging in GIS to identify the impact of fishing fleet for each of the 14 ports in the Catalan coast (Spain) to three demersal commercial fish resources. Soh *et al.* (2001) made an interesting comparison between current management systems and refuge management systems for shortraker and rougheye rockfish in the Gulf of Alaska. They used time series of catch data and a GIS model for designing harvest refugia networks of varying spatial extent. They showed that refugia can be used to greatly reduce discards and serial overfishing of stocks and substocks without reducing current catch levels, revealing the potential role of harvest refugia in fisheries management. Different scenarios of marine refugia-based management have been explored using also the SHADYS simulator (Maury and Gascuel 1999). SHADYS is a tool that interfaces numerical models with a GIS for the representation of spatial processes involved in fisheries dynamics. It is based on simple, realistic and well-identified fisheries related mechanisms, such as density dependent habitat selection, advection and diffusion of a fish population, fishermen search strategy, etc. It is shown that for diffusive or migratory species, the yield per recruit as a function of the protected marine surface can reach a maximum making the concept of 'space overfishing' meaningful.

### 3.3.2 Aquaculture

Aquaculture has received increased importance since its products aim to provide protein source for the increasing population worldwide, especially in developing countries. Identification and siting of aquaculture facilities is a complex spatial problem and requires expert knowledge of the marine and terrestrial environment as well as the understanding of several human factors. There are several criteria to assess aquaculture potential, including salinity, bathymetry, land uses, infrastructure, shelter and security, proximity to rivers or saltwater, and soils. For example, salinity extremes can greatly affect growth because raised fish populations are stationary and unable to move within the water column to find preferred combination of temperature and salinity ranges. Bathymetry can affect both the speed of the water (currents), which is important for fecal content and excess feed transport. Wave height is important for similar reasons as well as for cage security. Such factors can radically change the benthic environment due to increased nitrogen, which could lower the level of dissolved oxygen in the water. In certain cases, these factors might allow the raising of sea urchins, which are placed on the kelp below the cages thriving in the nitrogen rich environment. In addition, land use and soils are important because they could result in eutrophication or increased cloudiness of the water from erosion. Infrastructure also plays important role since aquaculture facilities should be easily accessible and at the same time be placed away from residential and tourist/leisure areas. Finally, shelter and security are important issues for the placement of aquaculture pens. For example, many wild predator species (e.g. seals) could attack aquaculture facilities (e.g. salmon pens). Placement of pens further offshore rather than in more protected areas (bays and inlets) results in exposing the cages and the fish to more stressful environments.

The use of GIS in the identification and siting of aquaculture facilities is highly suitable since it makes possible to identify locations that are favourable for the placement of aquaculture pens. Several studies integrate GIS for this purpose, providing a time effective and economically efficient way to analyse the parameters involved in choosing a site for aquaculture. Initial reported efforts include those by Kapetsky *et al.* (1987) on GIS and RS-based plans for aquaculture development in Costa Rica, Meaden (1987) for trout farms in UK, Kapetsky *et al.* (1988) for catfish farming, Kapetsky (1989) for aquaculture development in Johor state and Paw *et al.* (1992) in Lingayen Gulf (Philippines), Ali *et al.* (1991) for carp culture in Pakistan, and Ross *et al.* (1993) for site selection of salmonid cage culture. Meaden and Kapetsky (1991) discussed the potential and suitability of GIS and RS in aquaculture and inland fisheries while Kapetsky (1994) developed a strategic assessment of fish farming potential in Africa and McGowan *et al.* (1995) made a similar study for northeastern Massachusetts.

Several review studies excel the role of GIS in aquaculture planning (e.g. Kapetsky *et al.* 1987; Beveridge *et al.* 1994; Aguilar-Manjarrez and Ross 1993, 1995) describing the capacity of GIS to integrate various data (including remotely sensed data) for optimum site selection. Populus *et al.* (1995) discussed the use of RS data in association with field and land mapping data for the study of water quality (pigments and suspended matter) in relation to shrimp development in Java Sea (Indonesia). Use of satellite imagery (AVHRR) through GIS was also made by Habbane *et al.* (1997) in a study for optimum aquaculture zones in Baie des Chaleurs (East Canada). They combined satellite SST with *in situ* measurements, such as salinity, current speed and chlorophyll pigments to identify the space/time variability of surface water characteristics. The proposed optimum aquaculture zone in the area was characterised by periodic water mass fluctuations and high CHL pigment concentrations, providing good environmental conditions for high biological productivity. Ahmad (1997) discussed the use of GIS in aquaculture development in India, primarily for site selection purposes. Aguilar-Manjarrez (1998) presented a growth model, which is incorporated into a GIS, for producing estimations of aquaculture yield potential over the entire African continent. Extensive methodology and GIS routines are provided. White *et al.* (1998) integrated temperature and salinity data with land use and anthropogenic activity for two South Carolina coastal estuaries to study the infection and prevalence of a common protozoan pathogen of the oyster, which is highly associated with these parameters. Lee and Glover (1998) used GIS and a statistical software package to examine the effects of contaminated waters on shellfisheries of filter-feeding bivalve mollusks, such as oysters, clams, mussels and cockles. They related microbiological data obtained from shellfish monitoring to the type and size of sewage discharges in the vicinity of harvesting areas under commercial harvesting controls placed by the European Commission Shellfish Hygiene Directive (91/492/EEC). Congleton *et al.* (1999) integrated infrared aerial photographs and maps of bottom types, topography and bathymetry in a GIS to study how variations in habitat parameters affect success in shellfish mariculture as well as to identify potential sites for shellfish grow out. Soletchnik *et al.* (1999) used a GIS with data on growth, sexual maturation, survival rates and environmental variables to study summer mortality in oyster culture sites in Marennes Oleron Bay in French Atlantic coastline. Mortality was related to air temperature and food availability. Franklin (1999) presented a variety of personal computer-based GIS systems developed for marine and coastal managers by the Centre for Environment, Fisheries and Aquaculture Science (CEFAS), an internationally recognised Centre of Excellence for research, assessment and advice on fisheries management and environmental protection. Salam and Ross (1999, 2000) used GIS to model different comparative production scenarios for brackish and freshwater shrimp and fish aquaculture in SW Bangladesh, especially for optimising sites selection for development of shrimp and mud crab culture.

Bolte *et al.* (2000) presented POND, a decision support tool developed for analysis of pond aquaculture facilities through the use of a combination of simulation models and enterprise budgeting. A simulation framework provides a generic integrated simulation of data handling, time flow synchronisation and feature communication necessary for complex model-based decision support systems. The POND architecture includes a series of databases, a number of expert systems, models of pond ecosystems and various decision support features, such as assembling alternate management scenarios, economic analysis and data visualisation. A typical POND simulation includes the assembling of multiple ponds and fish lots, their management settings and projecting of changes in the simulated facility over time. Models from POND were introduced in complex GIS to evaluate aquaculture potential over large areas in Latin America (Kapetsky and Nath, 1997) and Africa (Aguilar-Manjarrez and Nath, 1998). Nath *et al.* (2000) identified that the deployment of GIS for spatial decision support in aquaculture is slow mainly due to lack of appreciation of the technology, limited understanding of GIS principles and associated methodology and inadequate organisational commitment to ensure continuity of spatial decision support tools. They analysed these constraints in-depth and presented several cases of GIS application in aquaculture. Arnold *et al.* (2000) described a GIS-based approach for identifying appropriate sites for aquaculture grow out, derived from work on hard clam aquaculture lease site selection in Florida. Doskeland and Hansen (2000) expelled the use of GIS for mapping areas where aquaculture sites can be located. They presented a conceptual GIS model, which can be used for locating sites for fish farms and for indicating, at a gross level, how many fish could be produced at a particular site based on area and recipient capacities. Gupta *et al.* (2000) described the use of GIS in aquaculture under a comprehensive coastal zone management plan for Gujarat. Brackish water aquaculture sites were selected using GIS techniques. Hassen and Prou (2001) developed a simple GIS model that accounts the spatial pattern in topography and land use for modelling non-point sources of dissolved inorganic nitrogen and orthophosphorus loadings from aquaculture activities to surface coastal water. Such loadings are usually influenced by water renewal rates and the spatial positions of constituent sources. Roy (2001) discussed the latest RS training and education needs in India mentioning current and future uses of high-resolution satellite imagery for aquaculture site selection.

### 3.3.3 Inland Fisheries

GIS applications in inland fisheries studies are also widespread, focusing on general management activities, such as monitoring of the environment, identification of potential acquisition areas and effects of environmental changes on reservoir, lake and river/stream fish populations. Rogers and Bergersen (1996) discussed the potential application of GIS in aquatic fisheries research, providing two simple GIS applications in both raster and vector formats that evaluate the use of vegetation and shoreline by the northern pike and largemouth bass in two Colorado reservoirs. This analysis was very effective in illustrating what types of questions can be addressed by either a raster or vector-based system. Keleher and Rahel (1996) discussed the potential impacts of global warming on habitat loss and overall salmonid abundance and distribution in the Rocky Mountain Region in Wyoming. Through the use of GIS technology, the authors combined various databases (e.g. air temperature, elevation, water temperature) to assess the potential impact of global warming on salmonid populations. Bernardo (1997) developed a GIS to monitor fish growth and health in the lake systems of Little Abitibi Provincial Park (Ontario) and evaluate future monitoring directions.

Fisher and Toepfer (1998) provided a comprehensive analysis of the status of GIS in American education as well as discussed key issues on the use of the technology and future applications in fisheries science. The study excelled in its discussion of GIS potential for prescriptive fisheries application in areas such as modelling and forecasting changes in aquatic habitats, estimating fish population abundances in unsampled areas, developing fishing sampling designs and integrating human population trends with biological aquatic habitat trends. Milner *et al.* (1998) developed HABSCORE, a system of salmonid stream habitat measurement and evaluation based on empirical models of fish density in relation to combinations of site and catchment features. The system is a habitat evaluation method able to explain the spatial component of variance in fish population data. Fisheries GIS applications, such as setting spawning targets and integrated catchment management planning are being developed through the US Fisheries River Habitat Inventory, which links HABSCORE to a national fisheries classification system. Belknap and Naiman (1998) developed a GIS using digital elevation models, digital hydrography and relative thermal stability data for wall base channels (detected by an airborne thermal infrared scanner) to automatically locate river flood plains and terraces, the geomorphic structures responsible for the creation and maintenance of wall base channels in rivers. Such structures are important to juvenile coho salmon for refuge from high flows and for rearing habitat.

Olson and Orr (1999) combined in GIS tree growth, fish and wildlife habitat, mass wasting and sedimentation data and hydrologic models in a decision analysis effort for long-term forest land planning based on multispecies habitat conservation plans. A GIS was developed for the freshwater salmon fishery in Scotland incorporating topographic, geology, land use, water quality monitoring and ecological field survey information. The GIS tool integrates stream slope and width estimates with stream locations, electrofishing survey data and local knowledge to define accessible and inaccessible stream areas for salmon fisheries. Analysis helps managers to prioritise survey effort, estimate natural spawning distribution and identify suitable areas for stocking with eggs or fry (Webb and Bacon 1999). O'Brien-White and Thomason (1999) used the expertise of a committee of fisheries biologists to evaluate fish habitat of the Edisto River Basin (South Carolina) using GIS. They used USGS 1:24 000 scale digital line graphs to assign values for data on species composition, predesignated protected areas, riparian habitat, dams and impoundments, ditches and water quality. They created combined class values for high, moderate and low-quality fish habitats. Hawks *et al.* (2000) used GIS to classify the Meramec River Basin in Missouri into strata based on ecoregion boundaries, watershed boundaries and stream order. The GIS was used by the Missouri Department of Conservation to make management decisions on potential acquisition areas, which were determined for each stratum based on species richness, habitat characteristics, percent of public land and number of human impacts, such as gravel and ore mining. Chiarandini (2000) demonstrated the use of GIS for the management of hydrographic and fishing resources in the mountain region of the But stream basin, located in northeastern Italy. Cresser *et al.* (2000) predicted river water quality, a requirement in river catchment management, through the development of a GIS-based empirical model, which uses geological data for the riparian zone within 50 m of the river, calculates the discharge contributions that passed through soil and predicts final run-off quality. Adams *et al.* (2001) mapped the potential invasible area of introduced fish into high elevation lakes to establish priorities in stocking and eradication efforts in relation to endangering the native lake and stream fauna. De Silva *et al.* (2001) used GIS to integrate fishery yield data, fishing intensity, landing size together with selected limnological data, such as conductivity and chlorophyll for nine tropical reservoirs in Sri Lanka. They established relations of fish yield to environmental parameters. Stoner *et al.* (2001) combined Generalised Additive Models (GAM) and GIS to analyse quasisynoptic seasonal beam trawl surveys, environmental parameters and sediment data for the development of habitat association for winter flounder in Navesink River/Sandy Hook Bay estuarine system in New Jersey. GAMs showed that the distribution of newly settled flounder in spring collections was associated with low temperature and high sediment organic content, placing them in deep, depositional environments while larger fish were associated with shallow depths, higher temperatures and presence of macroalgae. GIS generated maps for the probability of capturing fish of particular sizes showing two important centres of settlement in the system, connected to hydrographic conditions, to the rapid shifts of nursery locations in relation to fish size and to the dynamic nature of nursery habitats, which either expand, contract or shift position with changes in key environmental variables.

## 3.4 FISHERIES DATA SAMPLING METHODS

Fisheries monitoring through various sampling methods is essential for observation of stock conditions. Most nations sample two main kinds of fisheries data: Fisheries catch data are recorded through the establishment of statistical areas of various extend, which divide oceans in statistical 'rectangles' while fisheries landings are recorded in the major fishing ports and markets. Four such sampling systems for NE Atlantic, offshore NW Africa, SE Mediterranean and Lake

Victoria are shown in Figure 3.3. The basic fishery data collected through these worldwide fisheries statistical systems include generally the following: Catch and effort data (by fishery gear and fleet), total catch in number and weight (by species and fishery), effort statistics and fishing location, catch composition (by length, weight and sex), biological data (e.g. on age, growth, recruitment, and stock distribution) and research data (e.g. environmental, oceanographic and ecological factors affecting stock abundance). These time series spatiotemporal data are invaluable for fish stock monitoring and highly suitable for GIS analysis.

There are several mechanisms for the verification of these data, such as vessel monitoring systems (VMS), on-board observer programmes, vessel hardcopy reports and digital logbooks and port sampling. In 1999, the International Conference on Integrated Fisheries Monitoring was organised by the governments of Australia and Canada in cooperation with FAO and NOAA in Sydney, Australia (Nolan 1999). Twenty-six countries participated in the conference and delegates presented papers included the catch monitoring by fisheries observers in US and Canada (Karp and McEldery 1999) and New Zealand (France 1999), the use of electronic logbooks in the Canadian east coast scallop fishery (Matthews 1999), the VMS in the Gulf of Mexico shrimp fishery (Mejias 1999) and the control system of the Portuguese artisanal octopus fishery (Pereira 1999). In Greece, fisheries data sampling and verification is carried out by a network of fisheries officials spread out in 25 fisheries units all over the Aegean and Ionian seas (SE Mediterranean). The sampling network is supported by 25 electronic logging systems, which forward data in a central GIS database.

In United States, the National Marine Fisheries Service (NMFS) and its predecessor agencies began collecting fisheries landings data in 1880. Landings data were collected during surveys of a limited number of states and years between 1880 and 1951 while after that year, comprehensive surveys of all coastal states have been conducted. The collection of US commercial fisheries landings data is a joint state and federal responsibility. The cooperative State/Federal fishery data collection systems obtain landings data from state fishery, landing reports provided by seafood dealers, federal logbooks of fishery catch and effort and shipboard and portside interview and biological sampling of catches. State fishery agencies are usually the primary collectors of landings data but in some states, NMFS and state personnel cooperatively collect data. Survey methodology differs by state but NMFS makes supplemental surveys to ensure that data from different states and years are comparable. Statistics for each state represent a census of the volume and value of finfish and shellfish landed and sold at the dock rather than an expanded estimate of landings based on sampling data. Principal landing statistics consist of the pounds and vessel dollar value of landings identified by species, year, month, state, county, port and fishing gear. Most states get their landings data from seafood dealers, who submit monthly reports of the weight and value of landings by vessel. Recently, states are switching to mandatory trip reports to gather landings data. At the conclusion of every fishing trip, seafood dealers and fishermen indicate their landings by species on trip reports and record other data such as fishing effort and area fished.

The Forum Fisheries Committee (FFC) is the major forum for the development of fisheries policy in the Central and Western Pacific. It is made up of 16 member countries including: Australia, Cook Islands, Fiji, Kiribati, The

Republic of the Marshall Islands, The Federated States of Micronesia, Nauru, New Zealand, Niue, Palau, Papua New Guinea, Solomon Islands, Tonga, Tuvalu, Vanuatu and Samoa. A fully integrated GIS is used for the VMS. The VMS is an Australian Defence Force funded project that monitors over 1500 vessels located within the 16 individual Pacific countries' EEZs. Vessels are fitted with INMARSAT-C transponders to allow for real-time polling of vessel positions.

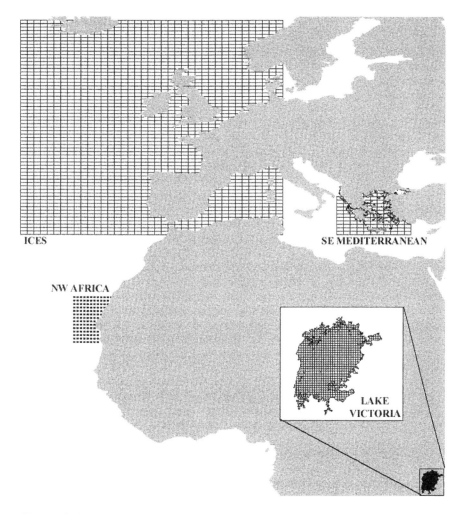

**Figure 3.3.** Various international and national spatial sampling schemes for the monitoring of fisheries production data in selected areas in Africa and Europe. Statistical rectangles are shown for NE Atlantic (ICES area), SE Mediterranean (Greece), off NW Africa (Senegal) and Lake Victoria (Kenya/Tanzania/Uganda).

Molenaar and Tsamenyi (2000) discussed international legal aspects related to satellite-based VMS for fisheries management presenting the significant legal restrictions, which exist in the exercise of jurisdiction by port and coastal states with respect to foreign vessels in lateral passage and conditions for entry into port. Kourti *et al.* (2001) presented preliminary results from a European Commission study on the feasibility of integrating SAR imagery as a tool for fisheries monitoring. They compared RADARSAT ScanSAR imagery to detected vessel positions with VMS position reports in the North Sea and the Azores area. The resolution of SAR imagery is sufficient to allow detection of vessels, thus providing a passive way to understand activity in an area for those vessels not using their VMS or not being subject to it or whose VMS is not working. Meaden and Kemp (1997a,b) identified that most of the world's commercial fishing fleets are equipped with manual data logging systems that allow for the collection of a range of data often with crude locational references. They presented FISHCAM, an Integrated Fisheries Computer-Aided Monitoring system, which is used on board fishing vessels and integrates both a GIS and a GPS with a relational database management system (http://www.cant.ac.uk/depts/acad/geography/fish/). The system allows analysis of catch and effort data in conjunction with other spatially related data (e.g. environmental). FISHCAM was developed at the Fisheries GIS Unit (Department of Geography, Canterbury Christ Church University College, UK) in collaboration with University of Kent's Computing Laboratory (Kemp and Meaden 1996). The system is composed of the on-board module, which is linked to a GPS and allows on-board digital recording of fisheries data and the management module, which is linked to a GIS database for monitoring purposes. FISHCAM is increasingly used in real-life situations, such as in monitoring shrimp resources off the coast of French Guiana. Among future applications of this software are the studies of normal range of fish stock and its fluctuations, relations between catches and local environmental indicators, relations between fish abundance and spatiotemporal variation of food supply and type, relations of fish protein indices and international needs, disturbance of benthic ecology by trawling, spatiotemporal variation of fish community structures, selection of location for fish enhancement projects (through fish biomass optimum environments) and monitoring of spatially related experimental fishing programmes.

Fox and Starr (1996) used GIS to compare Oregon trawl fishery time series catch data obtained from the Oregon Department of Fish and Wildlife (ODFW) and from the NMFS research cruises conducted at the same time in the same area. They concluded that information derived form ODFW logbooks can augment research data and improve estimates of the distribution and relative abundance of commercial fish species. Porter and Fisher (2000) discussed future directions in fisheries GIS underlining that advances in data collection technologies will increase the volume and types of associated data in the monitoring of commercial fisheries. They noted that there is considerable potential for using GIS in marketing recreational fisheries to segmented user groups. In the future, the potential of using GIS with fisheries monitoring data will definitely bring new automated methods in fisheries monitoring and data management.

## 3.5 FISHERIES DATA SOURCES AND GIS DATABASES

Several national and international fisheries databases provide information on commercial marine species in various scales. The Fisheries Department of the Food and Agricultural Organisation of the United Nations (FAO Fisheries) provides online worldwide statistical fisheries databases. The Southeast Asian Fisheries Development Centre (SEAFDEC) provides digital access to tables and figures on fisheries statistics in the area. The Sustainable Development Information Service of the World Resources Institute provides country level fishery statistics. The National Oceanic and Atmospheric Administration, Fisheries Statistics and Economics Division (NOAA Fisheries) provide fishery statistics at state level for US East Coast area. The International Council for the Exploitation of the Sea provides to ICES country members fishery statistics using a sampling grid system for NE Atlantic. In SE Mediterranean, fishery production is organised in GIS databases for data query and visualisation purposes. Sources of online fisheries data are listed in Table 3.2. These Internet sources of fisheries data include actual data, metadata or contact information for instruction on how to obtain or download data. Commonly, fisheries data are provided as tables or through online form based query search engines. In many cases, part of data are organised in GIS databases (e.g. for specific project purposes), thus part of fisheries statistical data are already appropriately organised for GIS applications. However such databases are not always easily accessible. Such fisheries GIS databases are usually developed for small area fisheries or include data for a short-time period, but contain detailed catch and landing information, extensive biological data and time series production data for many fishing gears (e.g. purse seiner, crabber, troller, trawler, longliner, gillnetter, etc.) providing a very good example of the capabilities of future nationwide, continuously updated similar databases.

**Table 3.2.** Internet sources of fisheries statistics and species life history data.

| ORGANISATION NAME | WEBSITE URL | NOTES |
|---|---|---|
| Department Of Fisheries, Malaysia Malaysia Fisheries | http://agrolink.moa.my/dof/dof1.html http://agrolink.moa.my/dof/statdof.html | Various summaries of annual fisheries statistics in Malaysian waters mainly for the period 1990–1996 |
| European Commission/EUROSTAT | http://europa.eu.int/comm/eurostat/ | Fishery statistics newsletter (Europe) |
| EC/MARSOURCE | http://europa.eu.int/comm/fisheries/policy_en.htm | European Common Fishery Policy |
| Falkland Islands Government Fisheries Department | http://fis.com/falklandfish/ | Annual catch and CPUE fisheries data for Falkland Islands EEZ (1998–2000) |
| Japan Information Network | http://jin.jcic.or.jp/stat/category_07.html | Japanese annual fisheries production (1985–1999) |
| European Union Agriculture and Fisheries | http://sunsite.nus.edu.sg/bibdb/pub/dsi/dsi004.html | Purchase information for the European Union Agriculture and Fisheries CDROM |
| SW Fisheries Science Center | http://swfsc.ucsd.edu/ | US commercial fisheries monthly and annual landings |
| SW Region's Sustainable Fisheries Division | http://swr.ucsd.edu/fmd/sustaina.htm | Aquaculture, landings and import/export data for US and Japan |
| Agriculture, Fisheries and Forestry, Australia | http://www.affa.gov.au/ | Australian various fisheries datasets |

| | | |
|---|---|---|
| Marine Biology Database | http://www.calpoly.edu/delta.html | Life history data for many marine species |
| Commission for the Conservation of Antarctic Marine Living Resources | Http://www.ccamlr.org/ | Instructions on how to download fisheries data for the Antarctic region |
| Cephbase | http://www.cephbase.dal.ca/ | An extensive cephalopod database |
| Alaska Commercial Fishing Data | http://www.cfec.state.ak.us/mnurpts.htm | Permit and fishing activity by year, state, census division or Alaskan city (1980–2000) |
| Fishbase | http://www.cgiar.org/iclarm/fishbase/ | An extensive fish database |
| Atlantic Coastal Zone Database Directory | http://www.dal.ca/aczisc/acdd | Metadata on 500 coastal and marine databases for Atlantic Canada |
| California Department of Fish and Game | http://www.dfg.ca.gov | Various data on US sport fishing |
| Canadian DFO Fisheries Statistics/Fisheries and Oceans Canada | http://www.dfo-mpo.gc.ca/communic/ statistics/stat_e.htm | Aquaculture and landing data for Canada |
| Marine and Coastal Management, South Africa | http://www.environment.gov.za/mcm/ | South African fisheries information |
| Europe Wide Fishing Ports Index | http://www.eurofishsales.com | Europe's premier website for fish landings and prices |
| NOAA/Alaska Fishery | http://www.fakr.noaa.gov/sustainable fisheries/catchstats.htm | Groundfish catch statistics and summaries for Alaska (1993–2001) |
| FAO Atlas of Tuna and Billfish | http://www.fao.org/fi/atlas/tunabill/ english/home.htm | Worldwide maps of tuna and billfish catches |
| FAO Fisheries Global Information System | http://www.fao.org/fi/figis/ | Extensive fisheries GIS from FAO |
| FAO Fisheries Department | http://www.fao.org/fi/statist/statist.asp | FAO fisheries and aquaculture statistical databases |
| FAO Fishery Country Profiles | http://www.fao.org/WAICENT/ FAOINFO/FISHERY/fcp/fcp.asp | Fishery Country Profiles with links to national fisheries data centres |
| FAO/ADRIAMED Project | http://www.faoadriamed.org/ | Fleet and catch information for the Adriatic Sea |
| US Federal Statistics | http://www.fedstats.gov | Gateway to statistics from over 100 US federal agencies |
| Ministry of Fisheries, New Zealand | http://www.fish.govt.nz/ | |
| Icelandic Ministry Of Fisheries | http://www.fisheries.is/ | Icelandic fisheries description by species |
| New England Fisheries Development Association | http://www.fishfacts.com/ | Extensive life history data for many marine fish species |
| FISHLINK | http://www.fishlink.com/html/ aquaindex.html | Annual European production data per country for carp, eels, salmon, sea bream and trout (1995–2000) |
| Directorate of Fisheries, Island | http://www.fiskistofa.is/dirfish/ | Catch data (1993–2001) and catch data in the Icelandic EEZ (1994–1999) |
| Florida Marine Research Institute | http://www.floridamarine.org/ | Life history data for many marine species |
| GDSourcing/Canadian statistics | http://www.gdsourcing.com/works/ Fisheries.htm | Various aquaculture and landing data for Canada |
| Great Lakes Fishery Commission | http://www.glfc.org/ | Commercial fish production in the Great Lakes (1867–2001) |
| British Columbia Fisheries | http://www.gov.bc.ca/fish/ | Annual catch and landings |
| Interior Columbia Basin Ecosystem Management Project | http://www.icbemp.gov/ | Databases relating to fish and fish habitat for the Interior Columbia Basin Ecosystem |
| International Commission for the Conservation of Atlantic Tunas | http://www.iccat.es/ | Software and database for annual fish catches of Atlantic tuna |
| International Council for the Exploitation of the Sea | http://www.ices.dk | Information on ICES fisheries data |
| Fisheries/Hong Kong | http://www.info.gov.hk/afd/fish/ | Information on aquaculture and marine fisheries in Hong Kong |

| | | |
|---|---|---|
| International Pacific Halibut Commission | http://www.iphc.washington.edu/halcom/commerci.htm | Commercial catch tables of Pacific halibut (1998–2000) |
| Fish Biodiversity in the Neotropics (NEODAT) | http://www.keil.ukans.edu/~neodat/ | A collection of databases on Neotropical Ichthyology |
| NOAA Library | http://www.lib.noaa.gov/china/chinafp.htm | General China fishery statistics |
| Ministry of Agriculture, Forestry and Fisheries of Japan | http://www.maff.go.jp/eindex.html | List of statistics on Japanese fisheries |
| Marine Life Information Network (UK and Ireland) | http://www.marlin.ac.uk/ | Extensive life history data on many marine species |
| North Atlantic Fisheries College | http://www.nafc.ac.uk/ | Various fisheries information for the Shetland Islands region |
| Northwest Atlantic Fisheries Organisation | http://www.nafo.ca/ | Fisheries data by country, species and division in NAFO areas (1994-2000) |
| National Biological Information Infrastructure | http://www.nbii.gov/ | Extensive worldwide biological information |
| North Carolina Division of Marine Fisheries | http://www.ncfisheries.net/statistics/ | Recreational and commercial fisheries catches in western Atlantic (1972–2000) |
| NE Fisheries Science Center (Woods Hole) | http://www.nefsclibrary.nmfs.gov/dbs.html http://www.nefsclibrary.nmfs.gov/land.html | List of worldwide fisheries databases |
| NMFS/Guide to Essential Fish Habitat Descriptions | http://www.nero.nmfs.gov/ro/doc/list.htm | Habitat information for many New England, Mid-Atlantic, Highly Migratory and South Atlantic species |
| NMFS Fisheries Statisitics | http://www.nmfs.noaa.gov/ | US commercial fisheries landings |
| NOAA Fisheries/Essential Fish Habitats | http://www.nmfs.noaa.gov/ess_fish_habitat.htm | General essential fish habitat information in US |
| NOAA Fishery Management Councils | http://www.noaa.gov/nmfs/councils.html | List of US Fishery Management Councils |
| Northwest Indian Fisheries Commission | http://www.nwifc.wa.gov/ | Salmon fisheries data for NW US |
| USGS/Index to Species Profiles: Life Histories and Environmental Requirements of Coastal Fishes and Invertebrates | http://www.nwrc.gov/publications/specindex.html | Life histories and environmental requirements of coastal fishes and invertebrates |
| Organisation for Economic Cooperation/Development | http://www.oecd.org/agr/fish/ | Review of Fisheries in OECD Countries: Vol I: Policies and Summary Statistics Vol II: Country Statistics 2000 Edition |
| Fisheries and Oceans Canada, Pacific Region | http://www.pac.dfo-mpo.gc.ca/English/default.htm | Fisheries management areas, Pacific Region |
| Pacific States Marine Fisheries Commission | http://www.psmfc.org | Fisheries data sampling manuals |
| Pacific Fisheries Information Network (PacFIN) | http://www.refm.noaa.gov | Maps of cumulative pollock fishery harvests and CPUE and flatfish CPUE for NE Pacific (1991-2000) |
| SE Asian Fisheries Development Center (SEAFDEC) | http://www.seafdec.org/data/ | Fishery Statistical Bulletin for the South China Sea (1991-1995) |
| Alaska Department of Fish and Game | http://www.state.ak.us/local/akpages/FISH.GAME/adfghome.htm | Extensive US fisheries information and statistics |
| NW Aquatic Information Network (STREAMNET) | http://www.streamnet.org/ | Fisheries information for waters within the Columbia Basin |
| USGS/Fisheries Data | http://www.umesc.usgs.gov/data_library/fisheries/fish_page.html | List of fisheries data servers in US |
| Fisheries Western Australia | http://www.wa.gov.au/westfish/ | Extensive fisheries information and statistics for western Australia |
| West Pacific Regional Fisheries Management Council | http://www.wpcouncil.org/ | Documents and fisheries maps for western Pacific |
| Sustainable Development Information Service | http://www.wri.org/sdis/ | General environmental data (including fisheries) from the World Resources Institute |

Some measures dealing with fisheries are difficult to quantify, especially for multispecies fisheries. Such measures depend on the high diversity in fishing vessel types and fishing gears and methods. Fishing Effort (FE) and Catch per Unit Effort (CPUE) are important factors in stock assessments, indicating trends in stock sizes and in exploitation rates. Generally, FE is normally defined as the product of Fishing Capacity (FC) and Fishing Activity (FA). FC (also called Fishing Power, FP) is a measure of the capability to catch fish and integrates the fishing vessel, fishing method and fishing crew. In FC calculation, vessel size, power and hold capacity as well as the size or quantity of fishing gears are important. FA indicates time at sea or number of fishing operations. Different measures of FA are applied to the various fishing methods. FA is usually measured in units of fishing time (e.g. hours fished, days on fishing grounds or days absent from port), number of fishing hauls or average number of set nets per week.

The georeferencing of various fisheries data into GIS databases provides many advantages for data quality checking, data update, query and visualisation as well as significant benefits to fishery scientists, resource managers and the fishing industry. In addition, the complex fisheries data sampling process is greatly facilitated when it is combined with GIS-connected digital data logging systems and automated GIS interfacing of fisheries catch data. Fisheries data are particularly suitable for relational database management systems and such systems are commonly used by many commercial and freely available GIS software packages. Durand (1996) and Valavanis *et al.* (1998) presented methods of introducing fisheries catch and landings data into GIS databases. These methods include simple conversion of ASCII tables to GIS relational tables, which are spatially linked to statistical 'rectangle' systems. Examples of fisheries catch and landings data in ICES and SE Mediterranean statistical systems are presented in Figure 3.4.

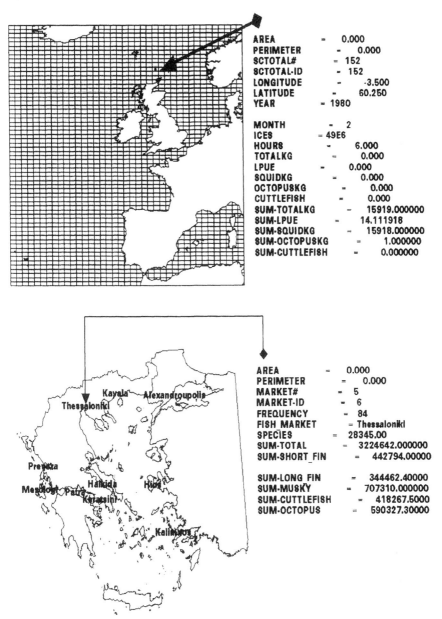

**Figure 3.4.** Organisation of fisheries catch and landing data into a GIS database. Each ICES statistical rectangle (above) and each Greek fish market (below) are linked to a relational database management system. ICES database is courtesy of Graham Pierce and Jianjun Wang, Zoology Department, University of Aberdeen, Scotland.

## 3.6 MAPPING PRODUCTION, BIOLOGICAL AND GENETIC DATA

The mapping of the various sampled fisheries data is a first step approach for revealing important spatiotemporal components of these data. In fisheries GIS applications, the mapping of fisheries production data is usually the initial step in data manipulation for further analysis. For example, Lee *et al.* (1999) used GIS to document the distribution patterns of three dominant tuna species (albacore, bigeye and yellowfin tuna) in the Indian Ocean based on catch data recorded by Taiwanese vessels, then investigated the characteristics of high abundance regions with SST distribution and CHL concentration and finally applied a discriminant function analysis to predict monthly distribution pattern using these catch and environmental variables. Also, spatial analyses of ocean policy, when integrated with fisheries production data, can provide an important component in balancing the conflicting uses of resources that occur in world oceans. Tools such as GIS help policy makers to identify gaps and overlaps in regulations. These types of decision support tools can lead to better management decisions and more integrated ocean management strategies. In order to conduct the necessary analyses, spatial deficiencies of policy and management regimes must be identified and addressed. New regulations must consider the state of the technology and adequately describe the geography under consideration and where possible, national agencies must clear up ambiguities in legal descriptions. For example, NOAA's Ocean Planning Information System (OPIS, http://www.csc.noaa.gov/opis/), a first attempt in the US to create a regional, multi state information system for the coastal ocean, is an application showing what can be accomplished when these spatial components are in place (Treml *et al.* 1999). Various methods of georeferencing legal frameworks were presented for southeastern US (Fowler *et al.* 1999; Fowler 1999), for various French metropolitan coasts (Guillaumont and Durand 1999), for administrative boundaries related to MPAs off the Spanish Atlantic coast (Sanz *et al.* 1999) and for administration of international maritime boundaries (Palmer 1998).

GIS abundance mapping was used by Gonzalez and Marin (1998) for the study of life cycle and abundance patterns of copepod species off the Chilean coast, which is dominated by the Humboldt Current system. They integrated in GIS species abundance, vertical (0–100 m) zooplankton samples and ancillary physical oceanographic data to show that species are abundant within 10 km from shore and that species remain year round in the upper water column with no evidence of seasonal vertical migration. Castillo *et al.* (1996) also used GIS techniques to analyse the distribution of three pelagic resources (anchovy, sardine and jack mackerel) off northern Chile. They used surveyed data on species catch and temperature/salinity distributions revealing that the distribution of the three species was associated with the occurrence and intensity of thermal and haline fronts. In a similar study, Abookire *et al.* (2000) used beach seine and small meshed beam trawl catch data as well as temperature and salinity data in nearshore and shallow habitats on the southern coast of Kachemak Bay (Cook Inlet, Alaska) to map fish distributions among habitats (based on species composition, catch and frequency of occurrence) in relation to differences in stratification. Macomber (1999) used GIS to map sole trawl logbook data from the Oregon trawl Fishery for the identification of species habitat spatial patterns.

The mapping of fisheries production through GIS includes several components: the geodistribution of catch per fish tool, landings per fish tool, fish tool activity areas (distribution of the fishing fleet from vessel monitoring systems or sampled major fishing fleet activity areas), fishing tool pressure on targeted species, major catch areas and species occurrence areas. A simple map overlay routine between the coastline and the catch values in each statistical rectangle of the fisheries sampling grid is used for the mapping of catch geodistribution (Figure 3.5). The same technique is applied between the coastline and landing values for the mapping of landings distribution in each fish market (Figure 3.6). Fishing tool pressure on targeted species may be shown in the form of graphs associating catch per fish tool data (Figure 3.7). Species occurrence areas are mapped by spatially integrating coastline/bathymetry and species catch geodistribution with constraint the species maximum depth of occurrence based on life history data. In addition, spatial integration of species occurrence areas and fishing fleet activity areas maps species major catch locations (Figure 3.8).

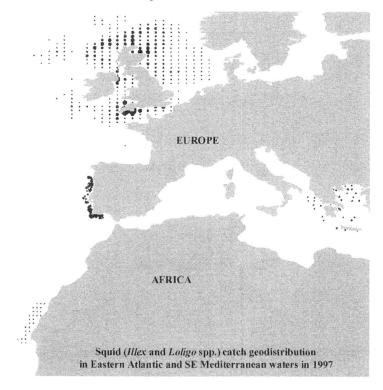

**Figure 3.5.** Synoptic GIS view of squid catch geodistribution in Eastern Atlantic and SE Mediterranean in 1997. Such views allow comparison of fisheries targeting similar species but occurring in different geographic areas. Data are courtesy of Graham Pierce and Jianjun Wand (University of Aberdeen, Scotland) for UK and Ireland, Joao Pereira (Instituto de Investigacao das Pescas e do Mar, Portugal) for offshore Portugal and Eduardo Balguerias (Centro Oceanografico de Canarias, Spain) for offshore NW Africa.

*Loligo vulgaris* **LANDING DISTRIBUTION 1996-99**

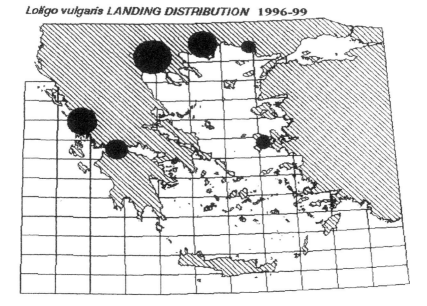

**Figure 3.6.** GIS mapping of fisheries landing geodistributions.

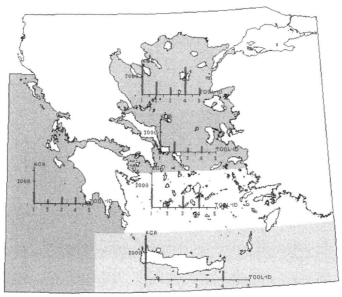

**Figure 3.7.** GIS output of fishing gear pressure on cephalopod populations in SE Mediterranean during the period January 1996 to December 1999. Fishing tool identification numbers are as followed: trawling (1), purse seiner (2), beach seiner (3), artisanal fisheries (4) and longliner (5).

*Loligo vulgaris MAJOR CATCH AREAS 1996-99*

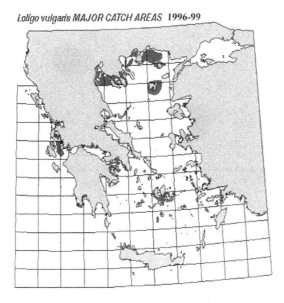

**Figure 3.8.** GIS output for major catch areas of common octopus in SE Mediterranean during the period January 1996 to December 1999. Extensive integrations among bathymetry, catch data, major fishing activity areas and species life history data (maximum depth of species occurrence) were performed for this GIS map output.

These simple GIS techniques reveal the spatial and temporal components of various fisheries data, which in turn, provide important information on species spatiotemporal distributions. Depending on species and the fishery, geodistributions of catch and landings may be either dispersed in wide areas or concentrated in specific smaller areas. These techniques may reveal the relation of catch and landings data and the association of catch areas to topographic features (e.g. bathymetry) and may associate catches with, for example, the edge of continental shelves or plateau areas. Depending on sampled fisheries data, these approaches provide information on the production potential of specific areas through specific fishing tools, which are invaluable to fisheries managers for the development of spatial components of management strategies (e.g. identification of seasonally overfished areas or areas proposed as marine reserves).

The mapping of sampled biological fisheries data (e.g. length, weight, sex, etc.) is very important in facilitating stock assessments as well as identifying overfished areas, especially when combined with catch/effort data. GIS applications incorporating such data help in the identification of areas where small fish are caught and areas of species recruitment. In addition, GIS mapping of genetic data facilitates the geographic definition of different stocks, interbreeding occurrences and the geographic extend of species diversity. Such data provide information on gene flow and give a general description of populations' genetic structure. Simple examples of biological and genetic data mapping within GIS are presented in Figures 3.9 to 3.12.

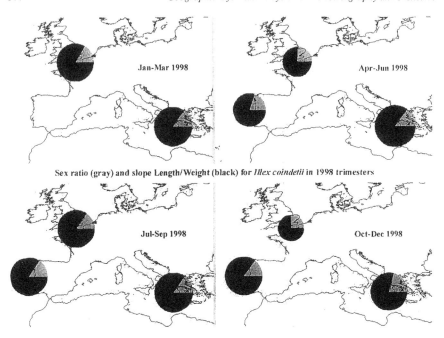

**Figure 3.9.** GIS mapping of biological data. Sex ratio (grey) and slope L/W (black) for a squid species in trimesters during 1998 fishing season. Data were obtained from fished squids in the English Channel, Portugal and SE Mediterranean.

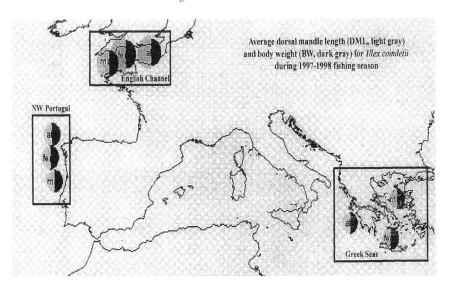

**Figure 3.10.** Average DML and BW for fished male, female and overall *Illex coindetti* during the 1997–1998 season in three major fishing areas in European waters.

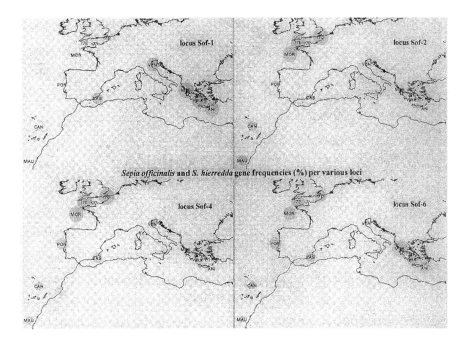

**Figure 3.11.** Percentages of gene frequencies per locus for two squid species in eastern Atlantic and Mediterranean waters.

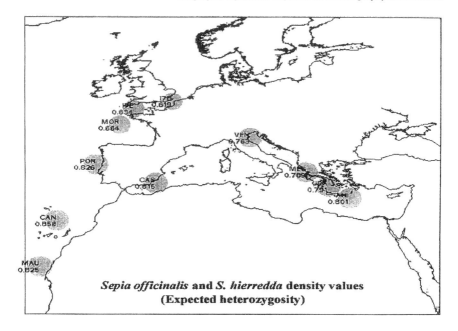

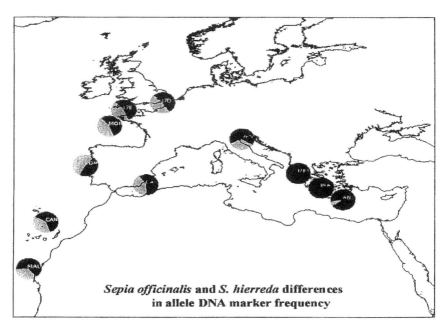

**Figure 3.12.** GIS mapping of expected heterozygosity and allele DNA marker size developed from samples of two squid species taken in various locations in eastern Atlantic and the Mediterranean.

## 3.7 MAPPING OF SPAWNING GROUNDS

Identification and mapping of species seasonal spawning grounds are of vital importance to species population regeneration, protection and management. Spawning areas are an important part of species essential habitats consisting a very sensitive area, which is often related to certain environmental variables (e.g. temperature and salinity), bottom substrate types and bathymetry. GIS map these sensitive areas by integrating several datasets that describe species preferred spawning conditions. A focal point in the process of mapping spawning grounds through GIS is the mapping of bottom sediment types, which was described earlier. The bottom substrate grid may be integrated with temperature distribution (SST), salinity (SSS), and bathymetry using life history data on species spawning preferences on SST, SSS, bathymetry ranges and substrate type as constraint parameters in these spatial integrations to reveal potential species spawning grounds.

Use of satellite imagery for the study of variation in spawning environmental conditions are used in several GIS studies towards identification of species spawning grounds. Lluch-Belda *et al.* (1991) identified sardine and anchovy spawning areas as these related to temperature and upwelling in the California current system. Simpson (1994) mentioned the use of high-resolution satellite imagery to spawning identification studies. Kiyofuji *et al.* (1998) analysed SST imagery to identify the variability of spawning ground distribution of Japanese common squid. Roberts (1998) developed a quantified model to study the influence of environmental parameters of chokka squid spawning aggregations. Waluda and Pierce (1998) used GIS to identify spawning locations of squids in UK waters. Geist and Dauble (1998) presented a conceptual spawning habitat model for fall chinook salmon that describes how geomorphic features of river channels create hydraulic processes, including hyporheic flows, that influence where salmon spawn in unconstrained reaches of large mainstem alluvial rivers. Quantitative measures of river channel morphology, including general descriptors of geomorphic features at different spatial scales, were incorporated in a GIS for the evaluation of substrate use and preference relative to available habitat. Van der Lingen (1999) used SST and data on the abundance and distribution of sardine and anchovy eggs in the southern Benguela upwelling region to define species spawning habitats. Varkentin *et al.* (1999) analysed temperature, plankton and current dynamics in western Kamchatka waters to identify the spawning characteristics of walleye pollock populations. Species spawning habits were related to warm or cold years and the invasion patterns of the West Kamchatkan current. Brown and Norcross (1999) used GIS and synoptic oceanographic data from satellite and ground measurements to define the spatiotemporal distribution of herring's early life history in Prince William Sound. Sakurai *et al.* (2000) analysed satellite SST data in GIS to infer distribution of spawning areas for *Todarodes pacificus* in East China Sea. The North Carolina Department of Environment, Health and Natural Resources, Division of Marine Fisheries in cooperation with the North Carolina Centre for Geographic Information and Analysis developed a digital GIS dataset for anadromous fish spawning areas in order to enhance planning, siting and impact analysis in areas directly affecting fish spawning. The digital dataset identifies locations of these sites in North Carolina (http://gis.enr.state.nc.us/datacatalog/).

Eastwood and Meaden (2000) developed a GIS model of spawning habitat suitability for sole in the eastern English Channel and southern North Sea using data on the distribution of sole eggs in relation to temperature, salinity, depth and sediment type. They accounted the importance of spatial variations in both habitat quality and quantity by using regression quantiles, a non-parametric regression technique that provides linear model estimates for any part of the biological response variable and therefore affords greater flexibility to modelling species/habitat relationships. Valavanis *et al.* (2002) integrated bottom sediment types, salinity, SST and bathymetry to identify cuttlefish and short-finned squid spawning grounds in SE Mediterranean. The method included integration of all those data applying integration constraints from species life history data (species preferred spawning environmental conditions). Figure 3.13 shows GIS derived spawning grounds for two species of squid in North Aegean Sea and Crete Island (Eastern Mediterranean). *Loligo vulgaris* spawning locations satisfy species spawning preferences of 10–25 °C during December-January, on hard substrate and depth up to 30 m. *Sepia officinalis* spawning locations satisfy species spawning preferences of 10–30 °C during March–July, on mud and rocks and depth up to 50 m.

SPAWNING AREAS IN NORTH AND SOUTH AEGEAN SEA

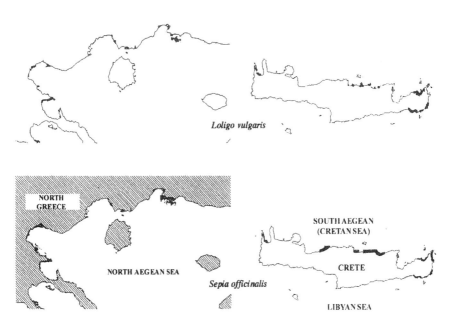

**Figure 3.13.** GIS output for spawning grounds of two cephalopod species in North and South Aegean Sea (SE Mediterranean). Extensive integration among substrate types, SST and salinity were performed with constraints taken from species life history data (preferred spawning conditions).

## 3.8 MAPPING ENVIRONMENTAL VARIATION OF PRODUCTION

The marine environment can influence the distribution of fish populations in at least two different ways. Water movement can move fish both horizontally and vertically and in various distances. Water temperature, oxygen content, proximity to land, nutrient and salinity extremes and other factors can either attract fish or force them away. The understanding of both influences is necessary to define suitable fish habitats, identify species preferred living environmental conditions and to forecast the abundance and distribution of fish stocks. Today, it is increasingly recognised that spatial and temporal trends in fish stock distribution and abundance may be related, at least in part, to environmental variation. Stock assessment studies on bigeye tuna, for example, the principal deepwater target species of the long line fishery in the Pacific Ocean, require the use of catch data as an index of the species abundance. Fishery dependent catch data (as it is the case for most commercially important species) does not necessarily represent the abundance of stock but rather the catchability of the stock. In turn, catchability is dependent, to a considerable extent, upon the variability of oceanographic conditions. It is oceanographic variability that can significantly affect the depth of the thermocline and the most likely depth of occurrence of bigeye tunas. Since it is known from biological studies that the preferred foraging habitat of bigeye tunas is the 8–15 °C water at or near the base of the thermocline, it is apparent that the importance of mapping the environmental variation of species production is essential for their management and protection to certain sensitive stages through their life cycle. Another example is that of squids and octopuses, which are species highly influenced by temperature and salinity. Again, the mapping of environmental variation in areas of species production provides valuable information on species preferred environmental conditions and seasonal habitats of aggregation.

Govoni (1993) studied larval fish distributions in the Gulf of Mexico in relation to the Mississippi river plume front and the western Gulf Stream front. Lee *et al.* (1994) studied the evolution of the Tortugas gyre and its influence on species recruitment in the Florida Keys. Drinkwater *et al.* (1994) used satellite imagery of SST to index the position of the Gulf Stream Front in northwest Atlantic. These indices that are associated with the North Atlantic Oscillation (weak winter NW winds) were updated in 1997. Dawe *et al.* (2000) integrated these indices with catch distribution of short-finned squid in the area offshore Newfoundland and found that squid abundance was positively related to favourable oceanographic conditions. In particular, the geographic distribution of squid populations was highly associated with that of the Gulf Stream meandering formations. Mathews (1999) studied the relations of small pelagic species distributions with several oceanographic processes (e.g. fronts, upwelling) in four Indonesian straits. Mokrin *et al.* (1999) studied the spatial distribution of flying squid in NW Japan Sea (Russian EEZ). Distribution, abundance and movements of the species were related to temperature, thremocline gradient and generally to water masses of different types of water structures. Demarcq (1999) presented CUSSI, a GIS-based application software for managing satellite data in coastal upwelling areas for fisheries management. The system integrates different satellite images to identify certain oceanographic processes and integrates results with fisheries catch data.

Zheng *et al.* (2001) applied a combination of principal components analysis (PCA) and cluster analysis to long-term average data for the definition of areas of similar seasonal patterns of whiting abundance in Scottish waters, based on fishery data on landings and effort. They used GIS to qualitatively describe the relationships of these spatial patterns of whiting abundance with trawl survey catch rates by age class and several environmental factors. Their results showed that the spatial patterns of whiting abundance are related to age, depth and spatial patterns of SST in winter.

Piatkowski *et al.* (2001) overviewed several studies that use GIS to identify cephalopods interaction with their environment. Such studies include the works by Waluda and Pierce (1998), who used GIS to map the temporal and spatial patterns in the distribution of squid *Loligo* spp. in UK waters and by Pierce *et al.* (1998), who mapped the distribution and abundance of *Loligo forbesi* in Scottish waters. Also, Du *et al.* (2000) used GIS to analyse a time series of mean weekly SST images and corresponding purse net statistic productivity for the period 1987–1997 in East China Sea. They found that SST data had great correlation with purse net productivity and their relationship varied steadily in certain range over time and area. Gaol and Manurung (2000) studied the effects of El Nino Southern Oscillation (ENSO) to tuna catches in South Java Sea (Indonesia). They combined bigeye tuna catch and time series of AVHRR SST imagery for 1996 non-ENSO year and 1998 ENSO year revealing that tuna catch during the ENSO year (lower temperatures) was higher than that of the non-ENSO year. Nakata *et al.* (2000) studied the implications of mesoscale eddies formed by fronts in the Kuroshio Current for anchovy recruitment. Waluda *et al.* (2001) used AVHRR SST images and catch data to study the interaction of squid *Illex argentinus* with the environment in SW Atlantic waters. Koutsoubas *et al.* (1998) and Valavanis *et al.* (2002) used GIS to identify similar interactions for five commercially important cephalopod species in SE Mediterranean. Santos *et al.* (2001) computed monthly SST upwelling indices along the Portuguese west coast using satellite images. Indices were related to data on sardine and horse mackerel recruitment dynamics to reveal that winter upwelling in the area corresponds to the spawning season for the species, thus having a negative impact on their recruitment due to an increase in those conditions that are favourable to the offshore transport of larvae and consequently an increase in their mortality.

Generally, the study of relations between species catch geodistribution and various environmental datasets may be approached in two main ways: First, the classification of surface waters (as described earlier) reveals distinct geographic areas with certain value ranges of oceanographic parameters (e.g. SST, salinity and CHL). A simple map overlay of the classification grid and species catch distribution shows the relation of species distribution to classified environmental parameters. Second, distributions of anomalies for certain environmental factors (e.g. SST and CHL) may be overlaid to species catch distribution and reveal relations of species distribution with the spatial range of environmental anomalies, which often consist a strong indication of seasonal front areas and possible upwelling regions. Figure 3.14 shows relations of *Illex argentinus* with surface temperature fronts in SW Atlantic and Figure 3.15 shows relations of total cephalopod catch with temperature anomalies in Eastern Mediterranean.

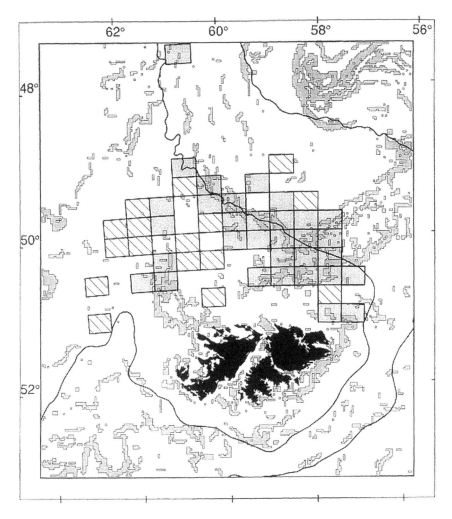

**Figure 3.14.** GIS overlay between *Illex argentinus* catch data (gray) and SST fronts in SW Atlantic fishing areas (Falkland Islands/Islas Malvinas). Species geodistribution is highly associated with SST fronts. Image is courtesy of Claire Waluda, British Antarctic Survey, UK (Waluda *et al*. 2001).

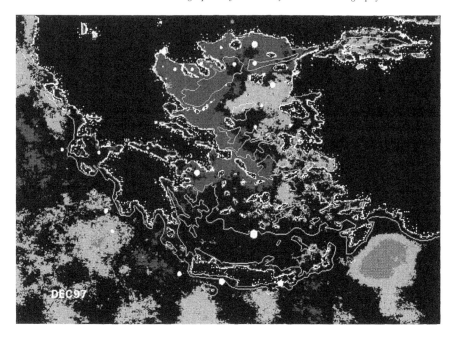

**Figure 3.15.** Cephalopod catch on satellite derived SST anomaly (December 1997, SE Mediterranean).
Cephalopod catches (white dots) are highly associated with boundaries of the geodistribution of SST
anomalies.

A wise managerial approach would be the development of oceanographic
conditions and variability atlases overlaid with historical and current fisheries
catch data in each EEZ or any area where commercial fisheries activities
commonly occur (Ramster 1994). Analysis of fisheries data is often complicated
by spatiotemporal limitations in the sampling programmes. Fisheries data are often
incomplete or certain areas are not adequately sampled. Thus, temporal analysis of
fisheries data should integrate spatial concepts. Use of GIS with extensive fisheries
and environmental data integration can potentially reveal hidden patterns and
relationships, invaluable for the development of protection and management
policies.

### 3.9 MAPPING MIGRATION CORRIDORS

Modelling of fisheries population movements (e.g. spawning or feeding migrations) using GIS is an application theme that has just recently got underway. The type of movement (horizontal or vertical) and its scale (small or large) play important roles in mapping species migration corridors (Schneider 1998). Another important factor, specifically related to spawning or feeding migrations of pelagic species, is environmental variation between aggregation areas and spawning or feeding grounds. This variation may be integrated in GIS migration modelling. For example, investigation of the migration routes of sardine (Tameishi *et al.* 1996), skipjack (Kawai and Sasaki, 1962), saury and mackerel schools (Saitoh, 1983) along the Japanese archipelago using acoustic aids (sardine relative abundance and chlorophyll concentration), satellite images (SST), aerial photographs (identification of sardine schools based on shape and colour) and commercial catch data showed that species utilise warm or cold streamers of the transition area of the Kuroshio and Oyashio currents during their inshore feeding migration. Rowell *et al.* (1985) showed several spatial components of squid populations in West Atlantic. They used commercial catches and SST to show how the distribution of the Gulf Stream affects the distribution of squid larvae and juvenile individuals as well as their migration patterns from Cape Hatteras to Newfoundland.

Movements of large predators may be also used as a factor for mapping fish migrations. In 1995, the Science Branch of the Fisheries and Oceans Canada (http://www.dal.ca) developed a GIS-based application with fishery and environmental data for integrated fishery management in the Newfoundland region. This pilot project integrated seal sightings with several environmental variables (temperature, ice distribution, bathymetry, etc.) into a GIS system. The resulted map layers provided a visual representation of seal distribution and environmental conditions suitable for developing hypotheses about the interactions of environment and seals as well as the basis for modelling these interactions particularly for the study of seal migrations and their forage areas. Also, Lowry *et al.* (1998) used the Argos system to monitor movements of tagged spotted seals in the Bering and Chukchi Seas (Alaska and Eastern Russia region). Yearly seal movements were associated with the seasonal extent of the location of the ice front with general southward movements during winter as sea ice coverage increased.

Fukushima (1994) identified the spawning migration of sakhalin taimen, a salmonid species, on northern Hokkaido Island. Isaak and Bjornn (1996) identified and related movements of squawfish and migrations of salmonids in Snake River. Arnold *et al.* (1994) studied movements of cod in relation to the tide generated streams in the southern North Sea while Schneider *et al.* (1999) studied similar movements of the same species in NW Atlantic. Kraus *et al.* (2000) studied the temporal and spatial variation of cod fecundity in the Baltic Sea. Denis and Robin (2001) created GIS tools for the analysis of spatial patterns in cuttlefish catches (French Atlantic fishery) from different fishing methods and related these patterns to landings per harbour. They showed that small spatial trends in catches are associated with species seasonal migrations in NE Atlantic.

Animal Movement is another recent effort for modelling spatial movements of fish populations. It is an ArcView GIS extension that contains a set of Avenue functions specifically designed to aid in the analysis of animal movement. The system was developed by USGS Alaska Biological Science Centre and the NFS Glacier Bay National Park and Preserve (Hooge *et al.* 1999; Hooge *et al.* 2000a; Hooge *et al.* 2000b). Animal Movement analysis is based on information on species habitat selection, relationships among individuals, population dispersion patterns or marine reserve efficacy and uses a set of classification tools and habitat selection algorithms.

An attempt to model a pelagic squid species' (*Loligo vulgaris*) offshore feeding and inshore mating and spawning migrations was made for the species' SE Mediterranean fishery (Valavanis *et al.* 2002). The developed GIS migration model is based on environmental variation factors and associated catch data. Temperature (satellite SST) and salinity integration with constraints from species life history data (species preferred minimum and maximum SST and salinity values) consist the main method. The species preferred ranges of SST and salinity were divided in three equally spaced groups: (1) 'Group 1' described ranges that were close to species preferred minimum values; (2) 'Group 2' described ranges close to average values; and (3) 'Group 3' described ranges close to maximum values. These groups of SST and salinity values were placed in a grid with three 'cost allocation' factors (1, 2 and 3), which revealed the 'difficulty' of a species to pass through a pixel based on species preferences in favourable environmental conditions (factor 2 being the most favourable). Finally, the model creates a path among adjacent cells that contain the average 'cost allocation' factor ('Group 2'). Migration corridor results were tested with offshore catch data, offshore upwelling and seasonal gyre formations in the region (Figure 3.16). In this example, *Loligo vulgaris* offshore/inshore migrations are related to species sexual maturity and occur during hunting. After copulation and spawning (March–July), first the males and then the surviving females migrate offshore to deeper regions and as far as 200 km from the coast. During winter (November–February), they migrate inshore for spawning.

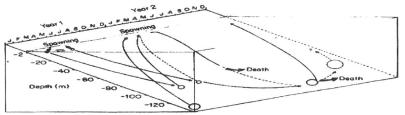

Diagrammatic view of Mediterranean migrations of *Loligo vulgaris*

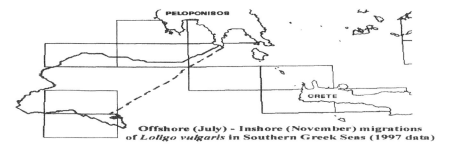

Offshore (July) – Inshore (November) migrations
of *Loligo vulgaris* in Southern Greek Seas (1997 data)

**Figure 3.16.** GIS modelling of seasonal offshore/inshore migration movements of *Loligo vulgaris*, an environmentally sensitive and highly mobile squid species. The image on the top is a diagram of species migrations in the Gulfe du Lion, French Mediterranean waters (Boyle 1983). Below, GIS modelled migration paths of offshore feeding (black line) and inshore mating/spawning (dashed line) migrations in the southern Greek Seas (SE Mediterranean) are shown with rectangles indicating presence of species catch data. The GIS migration model depends on seasonal environmental conditions and integrates species life history data (migration patterns and optimum living conditions), catch data and local oceanographic processes (e.g. gyres and offshore upwelling areas).

## 3.10 MAPPING SEASONAL ESSENTIAL HABITATS

Habitat is a general term to describe the physical, biological and ecological world of an organism. Marine species, in their full life cycle, cover certain geographical areas for their reproduction, recruitment, feeding and maturity. The seasonal mapping of such areas is essential for species monitoring and information-based management. In United States, the Magnuson/Stevens Fishery Conservation and Management Act of 1996 requires the regional Fishery Management Councils to describe and identify EFH for species under federal Fishery Management Plans. EFH are identified and described based on areas where species life stages commonly occur and include those waters and substrate necessary to fish to spawn, breed, feed and growth to maturity. To interpret the definition of EFH, 'waters' include aquatic areas and their associated physical, chemical, and biological properties that are used by fish, 'substrate' includes sediment, bottom structures underlying the waters and associated biological communities, 'necessary' means the habitat required to support a sustainable fishery and the managed species contribution to a healthy ecosystem and 'spawning, breeding, feeding or growth to maturity' covers a species full life cycle. To identify and describe EFH, the guidelines of the US National Marine Fisheries Service call for analysis of existing information at four levels of detail: (1) the presence and absence of distributional data of the geographic range of the species; (2) habitat-related densities of the species; (3) growth, reproduction and survival rates within habitats; and (4) production rates by habitat. Habitat Areas of Particular Concern (HAPC) are a subset of EFH. These are areas of particularly susceptible to human-induced degradation, ecologically important or located in an environmentally stressed region. In general, HAPC include high-value intertidal and estuarine habitats, offshore areas of high-habitat value or vertical relief and habitats used for migration, spawning and rearing of fish and shellfish.

Mapping of fish species habitats is very important for conservation efforts and similar to EFH concepts apply to other marine species, as well. For example, Moses and Finn (1997) integrated bathymetry, SST and right whale sightings data and developed a GIS-based logistic regression model to predict North Atlantic right whale distribution as a function of SST and bathymetry aiming to incorporate results into the Recovery Plan for the species. Laidre *et al.* (2001) estimated the carrying capacity for the California sea otter based on the density of sea otters at equilibrium within a portion of their existing range and the total area of available habitat. GIS was used to integrate observed numbers of sea otters, sediment and bathymetry data to classify substrates from the California coastline to the 40 m isobath as rocky, sandy or mixed habitat according to the amount of kelp and rocky substrate in the area and to finally map the potential species habitat.

Loneragan *et al.* (1998) tested the influence of seagrass characteristics on the distribution and abundance of post-larval and juvenile tiger prawns in the Gulf of Carpentaria (Australia) finding that the numbers of juvenile tiger prawns were lower in the low biomass seagrass beds and that these seagrass beds are the main nurseries for sustaining the production of the valuable Northern Prawn Fishery in Australia. Gaertner *et al.* (1999) used data from experimental trawl surveys in the Gulf of Lions (France) to study the spatial distribution of groundfish assemblages and to estimate their associations with benthic macrofauna and substratum types. Groundfish assemblages were split into two groups: (1) those strongly associated with both benthic macrofauna and type of substratum; and (2) those associated with the Rhone River plume and the shelf break upwelling. Logerwell and Smith (1999) analysed a 47-year dataset on the abundance and size distribution of larval sardine anchovy and hake. They used GIS to also integrate various oceanographic data and develop visualisations of areas of high-larval survivorship for these pelagic fish populations off California. Turk (1999) used GIS to obtain information on the life history of weathervane scallops in the Bering Sea. Integration of commercial logbook data, sediments and bathymetry showed that the species is highly correlated with sediment type and depth. Veisze and Karpov (1999) used GIS to present underwater video frames of marine organisms and their habitats captured from recordings made by a remotely operated vehicle (ROV) at the Punta Gorda Ecological Reserve (Humboldt County, California). Their development aimed to better observe the immense species richness and cryptic nature of many marine benthic organisms in the reserve. Wright *et al.* (2000) used underwater video observations and data on the abundance, sediment characteristics and depth to examine the physical characteristics of the habitat of the lesser sandeel in an attempt to predict the distribution of the species in North Sea, where sandeel is the subject of the largest single species fishery in the region. Guisan and Zimmermann (2000) reviewed the modelling efforts used for the prediction of species habitat distribution. Most approaches use ordinary multiple regression, neural networks, ordination and classification methods, Bayesian models, locally weighted approaches (e.g. GAMs) or combinations of these models. Valavanis *et al.* (2002) used extensive data selections and integrations in GIS to identify monthly squid suitable habitats in SE Mediterranean. The method included GIS integration of almost all available information on *Illex coindetii*, a highly mobile cephalopod species (Table 3.3). The first goal of these selections and integrations was to identify areas of potential species concentration and extract environmental conditions in these areas. Species concentration areas were considered as the common areas among species catch data, maximum occurrence depth and major fishing activity. The next goal was to use the values of the extracted environmental conditions on a monthly basis for identifying likely species preferred areas for each environmental variable (species life history data). The final GIS mapping of *Illex coindetii* seasonal suitable habitats was extracted by considering only these geographic areas where all species preferred environmental variables were present.

This environmentally oriented approach for mapping species suitable habitats suits to species that are sensitive to changes in environmental conditions and takes in account the majority of available data for the species (commercial catch data, fleet activity areas, life history data and the environment). In the case of *Illex coindetii*, the resulted GIS mapping of suitable habitats revealed the spatiotemporal distribution of species habitat, biology and migration habits (Figure 3.17). The fact that no areas of suitable habitat were identified during summer months (June, July and August) may be connected to species decreased growth rate from limited food supplies and species post-spawning high mortality. During fall and winter months, species growth rate increases and as a highly mobile and opportunistic species, they migrate offshore to take advantage of upwelling regions and associated plankton blooms. During spring months with spring spawning season approaching, species start their spawning migration in a southward direction to find warmer spawning and egg development temperature ranges.

A relatively similar environmentally oriented approach, which was further developed to a model was introduced by Huettmann and Diamond (2001) for environmental determination of seabird colony locations and distribution in Canadian North Atlantic. Their method includes integrated analysis of 20 marine environmental datasets, overlays with the PIROP database for pelagic seabirds, a Generalised Linear Model (GLM) for exploring the significance of environmental factors to seabird distribution and a Classification and Regression Tree (CART) for a detailed description of seabirds distributions. Such environmentally oriented approaches to modelling species habitats and geodistributions is highly suitable to species that are sensitive to changes in environmental conditions.

**Table 3.3.** List of GIS integrations among vector, raster datasets and species life history data for the final mapping of *Illex coindetii* predicted essential habitats.

| INTEGRATION DATASETS | GIS ANALYSIS TYPE | RESULT |
|---|---|---|
| 1. Species total catch coverage (rectangle system) | Selection for species catch more than 0 kg | Geodistribution of species catch |
| 2a. Geodistribution of species catch<br>2b. Species maximum depth of occurrence (bathymetric dataset and species life history data) | Spatial integration between polygon coverages | Geodistribution of species major occurrence areas |
| 3a. Geodistribution of species major occurrence areas<br>3b. Fishing activity areas | Spatial integration between polygon coverages | Geodistribution of species concentration areas |
| 4a. Geodistribution of species concentration areas<br>4b. Monthly SST, CHL, SSS Jan. image<br>…<br>4b. Monthly SST, CHL, SSS Dec. image | Spatial selection between a polygon coverage (vector) and an image (raster) | Minimum and maximum values of species SST, CHL, SSS preferences |
| 5a. Minimum and maximum values of species SST, CHL, SSS preferences per month<br>5b. SST, CHL, SSS monthly grids | Spatial selection in a grid using certain minimum and maximum values and conversion to polygons | Areas of species based on SST, CHL, SSS minimum and maximum values |
| FINAL INTEGRATION:<br>6a. Areas of species based on SST<br>6b. Areas of species based on CHL<br>6c. Areas of species based on SSS | Spatial integration among polygon coverages | Species predicted essential habitats on a monthly basis |

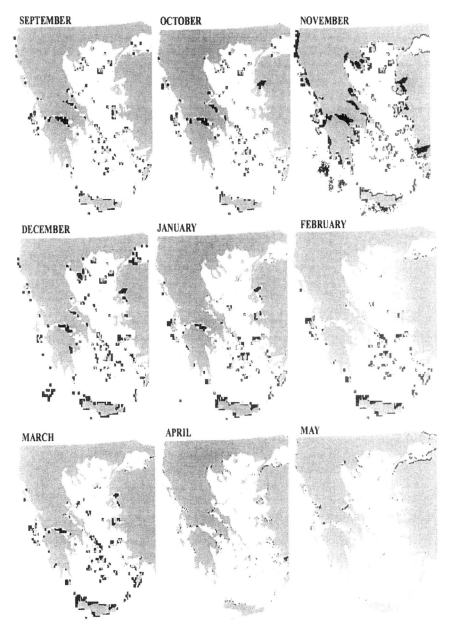

**Figure 3.17.** GIS output for monthly EFH for *Illex coindetii* in SE Mediterranean. Extensive data integrations were performed (see also Table 3.3) for the period 1996–1999.

## 3.11 GIS AND FISHERIES MANAGEMENT

Any natural resource is imbedded in an ecological, cultural and economic system. GIS tools can be used to manipulate spatial data that relate to any one of these aspects of resource dynamics. The knowledge of these dynamics is essential for resource management. Fisheries management is a multilevel and multidisciplinary process where marine scientists and policy makers develop management scenarios for the benefit of fisheries resources and fisheries communities. Fishing is an important economic activity worldwide. Fishing markets are large industries of highly significant source of employment, especially in areas where there are few alternatives. However, due to the large number of vessels for the exploitation of the existing stocks, there is usually an imbalance between imports and exports resulting in deficit in worldwide fishing markets. The high pressure of fishing vessels to fishing stocks leads to smaller stocks, smaller landings and smaller incomes. Overfishing is the main threat to the future of fish stocks and of the fishing industry. Especially in closed geographic areas, like the Mediterranean, where fishing pressure and the threat of pollution are high make conservation measures absolutely necessary. Today, sustainable fisheries worldwide are based on international conservation measures with the use of certain fishing techniques and methods, the fixing of some minimal net mesh sizes and sizes below which fish should not be landed.

The development of large-scale, information-based fisheries management is limited due to the lacking of monitoring and decision aid tools that integrate the capabilities of new technologies, such as GIS and RS. During the last 5 years, several countries have started developing GIS-based tools that analyse fisheries and RS data. However, the main aim of these tools is fisheries monitoring in a non-operational and/or research level by integration of various environmental and biological data for the study of relations among species population dynamics and oceanography. Results from these analyses are rarely integrated with the mapping of fisheries laws and socio-economic information. Mainly, this fact is due to the lack of technological development for the efficient mapping of fisheries laws.

In the global setting of the fishery sector, GIS fits in two main ways: (1) the monitoring of the fisheries components; and (2) the proposing of fisheries management schemes. Specifically, a marine fisheries GIS monitoring and decision aid tool must include the following components:

- The monitoring of fisheries production: To monitor fisheries yield production, a set of output should be targeted: the mapping of catch, landings, fishtool activity areas, species major catch areas, over and under fished areas, spawning areas, migration corridors and suitable habitats.
- The monitoring of oceanographic processes: To monitor oceanographic processes, the measurement and seasonalities of certain phenomena should be targeted: upwelling events, front systems, gyres as well as seasonal classification of surface waters according to their contents in temperature, CHL and salinity.
- The monitoring of fisheries socio-economic information: To monitor fisheries cultural and economic information, a set of outputs should be targeted: number

of fishing vessels, number of fishermen and number of fish product processing units.
• The mapping of fisheries laws and common fisheries policies: To check the efficiency of existing fisheries policies, fisheries laws must be spatially mapped and integrated with species population dynamics. This will result to the adjustment of current fisheries policies or the generation of new policies.

The main features of such a tool will be the automated data input, storage and analysis. Users should be able to register any data format to a GIS database, perform specific analytical tasks and reach spatiotemporal mapping outputs on:

• Relations between species population dynamics and oceanographic processes.
• Relations between species population dynamics and current fishery policies.
• Relations between species abundance and fisheries socio-economics.

These spatiotemporal relations will reveal over and under fished areas, map contradiction between species population dynamics and current fisheries policies and index various fishing ports based on the relation between fisheries socio-economics (e.g. number of anglers, power of fishing fleet) and targeted species populations.

Until recently, fisheries science has largely attempted to manage single species in isolation. It is well recognised that fisheries must be seen as part of a larger ecosystem and models of oceanic ecosystems should be developed and used in fisheries management. Since many fisheries are not sustainable, thus need rebuilding, large no-take regions should be established in fisheries areas. Refugia occur when environmental conditions act as a conservation mechanism to some or all of the species located in a particular area (Caddy 1984). These natural refugia are used by species several times through their life cycles, especially during mating and spawning. The identification and mapping of these areas could be used as an indirect approach to the regulation of fishing effort. In this mapping effort, many parameters play significant role, such as adundance geodistribution and fishing fleet monitoring as well as analysis of environmental processes. Marine reserves (as 'artifical' refugia) is a common management practice, however, from 71 per cent of ocean surface less than 1 per cent of the marine environment is within protected and marine reserve areas. All reserve designs must be guided by analytical information on natural history and habitat variability. For example, many pelagic species use predictable habitats to breed and forage. The MPAs could be designed to protect these foraging and breeding aggregations. The identification of the physical mechanisms that influence the formation and persistence of these aggregations is essential to define and implement pelagic protected areas (Hyrenbach *et al*. 2000). Murray *et al*. (2000) identified that MPAs are becoming a popular tool in fisheries management to prevent overexploitation of fish stocks (Polacheck 1990), to conserve biodiversity (Bohnsack 1993) and to reduce bycatch of non-targeted species (Parsons 1993; Mikol 1999). Though empirical support for MPAs is rising (Russ and Alcala 1996; Murawski *et al*. 2000), debate still exists over whether protected areas can effectively meet their objectives (Crowder *et al*. 2000). The use of GIS in fisheries may benefit researchers as they examine the spatial effects of MPAs and evaluate new and old management actions. A useful technique would be the seasonal classification of pelagic habitats according to their dynamics and predictability into three categories: (1) static; (2) persistent; and (3) ephemeral. This classification may be also applied to coastal habitats, which are often characterised by dynamic boundaries and extensive buffers. Identification and measurement of oceanographic features in various scales could be used for the design of dynamic MPAs, whose extent and location could be bonded to open ocean or coastal oceanographic processes. Based on these approaches, Baumann *et al.* (1998) developed a GIS database of coastal and marine protected areas, conservation zones and restricted fishing areas in the Gulf of Maine. Such systems consist valuable management tools because their flexibility provides the means for data update and seasonal reorganisation of the spatial boundaries of coastal restriction zones.

Another fisheries management issue is the reduction of problematic bycatches of non-targeted species. As identified at the International Conference on Integrated Fisheries Monitoring (February 1999, Sydney Australia), data on unwanted bycatches is limited. Fisheries monitoring systems should include on-board observers for the sampling of bycatches (Kennelly 1999). On this issue, Murray *et al.* (2000) introduced a detailed GIS approach including the management perspective. They conducted a GIS analysis to evaluate the effectiveness of a time/area closure designed to reduce the bycatch of harbor porpoises in the New England sink gillnet fishery. A GIS database was built to analyse bycatch rates before, during, and after certain closure periods and to track vessels in order to examine how fishermen responded to the closure. Analyses were made in two main steps. Bycatch Analysis included data provided by on-board observers on position, presence or absence of harbor porpoise bycatch, soak duration, number of nets in the string, depth at which the string was set, mesh size and target species. These data allowed for the calculation of bycatch rates and fishing effort inside and outside closure areas. Homeport Analysis included the calculation of the proportions of traditional fishing grounds of each adjacent homeport that overlapped the closure areas and the proportion of traditional fishing grounds that were off limits to fishermen from different homeports during the closure period. Through these GIS analytical steps, it was concluded that the effectiveness of time/area closures was correlated to the spatiotemporal variation in patterns of bycatch rates of harbour porpoises and that small time/area closures may have disproportionate impacts on fishermen involved (Figure 3.18).

Improvements in the design of fishing vessels, gears, navigation and fishing finding equipment has created highly efficient fleets with catching capacity exceeding the available resources. Today, in worldwide fisheries management efforts, fishing fleet activity is controlled by restricting access to the fisheries, limiting fishing effort and regulating catches. Free access is no longer possible and in many cases, no new fishing licenses are being issued. Fishing effort is regulated by restricting days at sea. Stock assessments indicate limits of total allowable catches and fish minimum landing sizes. Minimum mesh sizes are set for groups of species and cod end designs greatly improve the size selectivity of the fishing gears. Permanent, seasonal and fisheries specific area closures are also measures for the prevention of the destruction of concentrations of young fish. All these measures share at least on common component, that of the spatiotemporal dynamics of fish populations. GIS technology can offer the analytical results that can help fisheries managers to regularly monitor the dynamics of species populations, thus further improving the effectiveness of all fishing regulation measures.

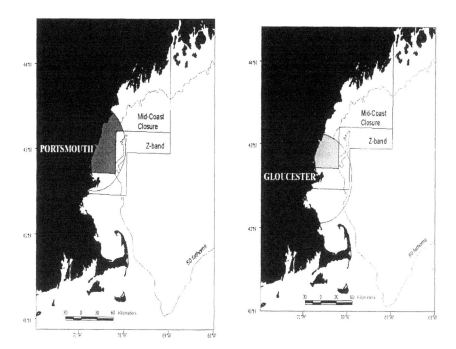

**Figure 3.18.** Marine protected areas (MPA) are becoming a popular tool in fisheries management to prevent overexploitation of fish stocks. However, debate still exists over whether protected areas can effectively meet their objectives. The use of GIS in the fisheries realm may grow as researchers examine the spatial effects of MPAs and evaluate new and old management actions. For example, the US mid-coast closure, designed to reduce the bycatch of harbor popoises in the northeast gillnet fishery, has disproportionate impacts on fishermen depending on their homeport. Sixty four per cent of the traditional fishing grounds of fishermen from Portsmouth, New Hampshire were eliminated by the mid-coast closure (shaded area left), versus a 25 per cent loss by fishermen from Gloucester, Massachusetts (shaded area right). Figures are courtesy of Kimberly Murray, Woods Hole Oceanographic Institute, Marine Policy Center, Woods Hole, MA (http://www.whoi.edu/mpcweb/).

Caddy (1999) mentioned that a move to economically efficient small-scale coastal fisheries, as these are integrated into wide coastal management schemes, will reinforce the need for local access controls and zonation of activities. This trend is already evident with GIS technology, which is used to categorise the suitability of spatial indices among and within coastal activities. GIS that are applied to integrated coastal resource management give coastal nations a range of geographically differentiated tools for dealing with access rights. Geographical differentiation of user rights is a management idea that may be applied in the near future. Also, a relatively recent trend in fisheries management is the consideration of the impacts of fishing on the structure and functioning of marine ecosystems, in addition to standard considerations focusing on the sustainable yield of the target species. FAO Code of Conduct for Responsible Fisheries (FAO 1995) reiterates the obligation of nations to consider the impacts of their environmental policies on marine ecosystems. Gislason *et al.* (2000) overviewed issues and developments towards incorporating ecosystem objectives to the conservation component of fisheries management as well as that of other ocean use fields. They underlined that the aggregated ocean use activities need to be examined in a nested manner based on a range of geographic scales as these related to broad definitions of conservation objectives. The spatial definition of any conservation objectives often becomes a difficult task, since the geographical distributions of marine populations vary enormously, as do the geographic scales of the processes and interactions that maintain populations in an ecosystem. In turn, integration of data in such spatially defined management units becomes a necessity, since the interaction of species populations and the marine environment is often very descriptive. In such a setting, fisheries monitoring, marine RS and marine GIS provide the data and the means for greatly facilitating management objectives in the scope of meeting conservation goals in an integrated technological environment.

## 3.12 SUMMARY

Applications of GIS technology in Fisheries are well established and growing. A variety of remotely sensed, surveyed, statistical and species life history data can be integrated through extensive GIS analysis resulting in the seasonal mapping of species population dynamics. This dynamic GIS mapping provides valuable information for fisheries managers, who continuously require background information for developing management scenarios.

The spatial nature of fisheries problems is well-recognised making GIS technology, the spatiotemporal data management framework, an invaluable analytical tool for the resolution of such problems. Today, spatial components of fisheries management, such as essential fisheries habitats and marine protected areas, are fully understood and in many cases are established by laws. The unique capability of GIS to integrate multidisciplinary datasets, which become a requirement for the complex fisheries management process, brings a new approach in the field, as GIS technology itself is a new technology. This attribute provides wide managerial approaches, which are supported by the integration of another two new technologies, those of RS and Global Positioning System and a series of other disciplines, such as scientific visualisation, geostatistics, image processing, etc. Adding monitoring, biological, genetic and socio-economic data, GIS technology brings under the same framework all available fisheries information as well as all available data analysis tools towards information-based fisheries management.

Currently GIS, as a new technology that is under ongoing development, imposes several technological problems in its application in fisheries. The resolution of 3D visualisation made a quite remarkable progress, however, fish populations, living in a highly dynamic 3D environment, make the ability of data integration in a 3D space a necessity and requirement for producing better models of marine dynamics. This capacity is currently not developed and as in the case of oceanographic GIS, the development of marine-related GIS software with this capability should be included in the production of newer versions of existing GIS packages or of new GIS software products. This development will definitely boost our current modelling GIS capabilities.

Major organisations and national fisheries data holding centres already use GIS as a tool for the organisation of their data and for the production of stand alone or online fisheries GIS applications. Signs show that in the near future, as technology progresses, marine and fisheries GIS methods and applications will become even more integrated and sophisticated than they are today, greatly facilitating the spatiotemporal aspect of fisheries and fisheries management.

## 3.13 REFERENCES

Adams, S.B., Frissell, C.A. and Rieman, B.E. (2001). Geography of invasion in mountain streams: consequences of headwater lake fish introductions. *Ecosystems*, **4**, 296–307.

Abookire, A.A., Piatt, J.F. and Robards, M.D. (2000). Nearshore fish distributions in an Alaskan estuary in relation to stratification, temperature and salinity. *Estuarine, Coastal and Shelf Science*, **51**, 45–59.

Aguilar-Manjarrez, J. (1998). GIS analyses for the assessment of fish farming potential in Africa. In J. Aguilar-Manjarrez and S.S. Nath, eds. *Strategic Reasssement of Fish Farming Potential in Africa*. CIFA Technical Paper, No. 32. FAO, Rome, 1998. http://www.fao.org/docrep/W8522E/W8522E10.htm

Aguilar-Manjarrez, J. and Nath, S.S. (1998). A strategic reassessment of fish farming potential in Africa. CIFA Technical Paper No. 32, p. 170, FAO, Rome.

Aguilar-Manjarrez, J. and Ross, L.G. (1993). Aquaculture development and GIS. *GIS Europe*, **7(4)**, 49–52.

Aguilar-Manjarrez, J. and Ross, L.G. (1995). Managing aquaculture development: the role of GIS in environmental studies for aquaculture. *GIS World*, **8(3)**, 52–56.

Ahmad, S.H. (1997). Role of GIS in aquaculture. In Misra B, ed. *Geographic Information System and Economic Development : Conceptual Applications*, No. 12485, p. 186, International Development Research Centre, Canada.

Ali, C.Q., Ross, L.G. and Beveridge, M.C.M. (1991). Microcomputer spreadsheets for the implementation of GIS in aquaculture: a case study on carp in Pakistan. *Aquaculture*, **92(2–3)**, 199–205.

Arnold, G.P., Walker, M.G., Emerson, L.S. and Holford, B.H. (1994). Movements of cod (*Gadus morhua* L.) in relation to the tidal streams in the southern North Sea. *ICES Journal of Marine Science*, **51(2)**, 207–232.

Arnold, W.S., White, M.W., Norris, H.A. and Berrigan, M.E. (2000). Hard clam (*Mercenaria* spp.) aquaculture in Florida, USA: geographic information system applications to lease site selection. *Aquacultural Engineering*, **23(1–3)**, 203–231.

Ault, J.S., Luo, J., Smith, S.G., Serafy, J.E., Wang, J.D., Humston, R. and Diaz, G.A. (1999). A spatial dynamic multistock production model. *Canadian Journal of Fisheries and Aquatic Sciences*, **56**, 4–25.

Baumann, C., Brody, S., Fenton, D., Nicholson, B. (1998). A GIS database of existing coastal and marine protected areas, conservation zones, and restricted fishing areas in the Gulf of Maine. Gulf of Maine Marine Protected Areas Project Report, p. 20.

Begg, G.A., Hare, J.A. and Sheehan, D.D. (1999). The role of life history parameters as indicators of stock structure. *Fisheries Research*, **43**, 141–163.

Bernardo, G.J. (1997). GIS unlocks the mystery of walleye movements. *GIS World*, **10(10)**, 42–47.

Belknap, W. and Naiman, R.J. (1998). A GIS and TIR procedure to detect and map wall base channels in Western Washington. *Journal of Environmental Management*, **52(2)**, 147–160.

Beveridge, M.C.M., Ross, L.G. and Mendoza, Q.M. (1994). Geographical Information Systems (GIS) for coastal aquaculture site selection and planning. In K. Koops, ed. *Ecology of Marine Aquaculture*, pp. 26–47.

Bez, N. and Rivoirard, J. (2000). On the role of sea surface temperature on the spatial distribution of early stages of mackerel using inertiograms. *ICES Journal of Marine Science*, **57**, 383–392.

Bohnsack, J.A. (1993). Marine reserves: they enhance fisheries, reduce conflicts, and protect resources. *Oceanus*, **4**, 63–71.

Bolte, J., Nath, S. and Ernst, D. (2000). Development of decision support tools for aquaculture: the POND experience. *Aquacultural Engineering*, **23(1–3)**, 103–119.

Booth, A.J. (1998). Spatial analysis of fish distribution and abundance: a GIS approach. In F. Funk, T.J. Quinn II, J. Heifetz, J.N. Ianelli, J.E. Powers, J.F. Schweigert, P.J. Sullivan and C.I. Zhang, eds. *Fisheries Stock Assessment Models*, pp. 719–740. Alaska Sea Grant College Programme Report No. AK/SG/98/01, University of Alaska Fairbanks.

Booth, A.J. (2000). Incorporating the spatial component of fisheries data into stock assessment models. *ICES Journal of Marine Science*, **57(4)**, 858–865.

Boyle, P.R. (1983). *Cephalopod Life Cycles. Species Accounts*. Vol. I. Academic Press, London.

Brown, E.D. and Norcross, B.L. (1999). Effect of herring egg distribution and ecology on year class strength and adult distribution. In *Proceedings of the 17th Lowell Wakefield Fisheries Symposium: Spatial Processes and Management of Fish Populations*, October 1999, Anchorage, Alaska.

Brown, S.K., Buja, K.R., Jury, S.H., Monaco, M.E. and Banner, A. (2000). Habitat suitability index models for eight fish and invertebrate species in Casco and Sheepscot Bays, Maine. *North American Journal of Fisheries Management*, **20(2)**, 408–435.

Caddy, J.F. (1984). An alternative to equilibrium theory for management of fisheries. *FAO Fisheries Reports*, **289(2)**, 173–208.

Caddy, J.F. (1999). Fisheries management in the twenty first century: will new paradigms apply? *Reviews in Fish Biology and Fisheries*, **9**, 1–43.

Caddy, J.F. and Carocci, F. (1999). The spatial allocation of fishing intensity by port based inshore fleets: A GIS application. *ICES Journal of Marine Science*, **56(3)**, 388–403.

Caddy, J.F., Refk, R. and Do-Chi, T. (1995). Productivity estimates for the Mediterranean: evidence of accelerating ecological change. *Ocean and Coastal Management*, **26(1)**, 1–18.

Caiaffa, E. (2000). European marine information system: Eumaris prototype. *Oceans Conference Record (IEEE)*, **3**, 1569–1575.

Carocci, F. and Majkowski, J. (1996). Pacific tunas and billfishes. Atlas of commercial catches, pp. 9, 28 maps. Rome, FAO.

Carocci, F. and Majkowski, J. (2001). Atlas of tuna and billfish catches. FAO, Rome. On line: http://www.fao.org/fi/atlas/tunabill/english/index.htm

Castillo, J., Barbieri, M.A., Gonzalez, A. (1996). Relationships between sea surface temperature, salinity, and pelagic fish distribution off northern Chile. *ICES Journal of Marine Science*, **53(2)**, 139–146.

Chiarandini, A. (2000). A geographical information system to manage water resources and fishing fauna. *Geo Informations Systeme*, **13(4)**, 14–20.

Congleton, W.R., Pearce, B.R., Parker, M.R. and Beal, B.F. (1999). Mariculture siting: a GIS description of intertidal areas. *Ecological Modelling*, **116(1)**, 63–75.

Cresser, M.S., Smart, R., Billett, M.F., Soulsby, C., Neal, C., Wade, A., Langan, S. and Edwards, A.C. (2000). Modelling water chemistry for a major Scottish river from catchment attributes. *Journal of Applied Ecology*, **37(1)**, 171–184.

Crowder, L.B., Lyman, S.J., Figueira, W.F., Priddy, J. (2000). Source sink population dynamics and the problem of siting marine reserves. *Bulletin of Marine Science*, **66(3)**, 799–820.

Dawe, E.G., Colbourne, E.B. and Drinkwater, K.F. (2000). Environmental effects on recruitment of short finned squid (*Illex illecebrosus*). *ICES Journal of Marine Science*, **57**, 1002–1013.

De Leiva Moreno, J.I., Agostini, V.N., Caddy, J.F., Carocci, F. (2000). Is the pelagic/demersal ratio from fishery landings a useful proxy for nutrient availability? A preliminary data exploration for the semienclosed seas around Europe. *ICES Journal of Marine Science*, **57(4)**, 1091–1102.

Demarcq, H. (1999). CUSSI: a software for managing satellite data in coastal upwelling areas for fisheries management in a GIS context. In *Proceedings of the First International Symposium on Geographic Information Systems (GIS) in Fishery Science*. T. Nishida, P.J. Kailola and C.E. Hollingworth, eds. Fishery GIS Research Group, Saitama, Japan.

Denis, V. and Robin, J.P. (2001). Present status of the French Atlantic fishery for cuttlefish (*Sepia officinalis*). *Fisheries Research*, **52(1–2)**, 11–22.

De Silva, S.S., Amarasinghe, U.S., Nissanka, C., Wijesooriya, W.A. and Fernando, M.J. (2001). Use of geographical information systems as a tool for predicting fish yield in tropical reservoirs: case study on Sri Lankan reservoirs. *Fisheries Management and Ecology*, **8(1)**, 47–60.

Doskeland, I. and Hansen, P.K. (2000). Geographic information systems (GIS) are tools for better integrated coastal zone planning and management (ICZP/M). In *Proceedings of ICES 2000 Annual Science Conference*, Theme Session on Sustainable Aquaculture Development, September 2000, Bruges, Belgium.

Drinkwater, K.F., Myers, R.A., Pettipas, R.G. and Wright, T.L. (1994). Climatic data for the Northwest Atlantic: the position of the shelf/slope front and the northern boundary of the Gulf Stream between 50 W and 75 W, 1973–1992. *Canadian Data Report of Hydrography and Ocean Sciences*, **125**, 1–103.

Drury, J. (1999). Mapping and classification of marine habitats. *Hydro International*, **3(6)**, 6–7.

Du, Y., Zhou, C.H., Shao, Q. and Su, F. (2000). Sea surface temperature and purse net productivity in East China Sea. *International Geoscience and Remote Sensing Symposium (IGARSS)*, **5**, 1872–1874.

Durand, C. (1996). Presentation de Patelle: Programmemation graphique de chaines de traitement SIG sous ArcView. IFREMER Note Technique TC/04. Plouzane: Sillage.

Eastwood, P.D. and Meaden, G.J. (2000). Spatial modelling of spawning habitat suitability for the sole (*Solea solea* L.) in the eastern English Channel and southern North Sea. In *Proceedings of ICES 2000 Annual Science Conference*, 27–30 September 2000, 88th Statutory Meeting, 24 September to 4 October 2000, Brugge (Bruges) Belgium. ICES CM 2000/N:05.

Fairweather, T., Booth, A. and Sauer, W. (1999). The development of a Fisheries Information System (FIS) to aid fisheries off the west coast of South Africa. In *Proceedings of the 17th Lowell Wakefield Fisheries Symposium: Spatial Processes and Management of Fish Populations*, October 1999, Anchorage, Alaska.

FAO (1995). Code of Conduct for Responsible Fisheries, p. 41, FAO, Rome. On line: http://eelink.net/~asilwildlife/FAOCodeofConduct.htm

Fisher, W.L. and Toepfer, C.S. (1998). Recent trends in geographic information systems educations and fisheries research applications at US universities. *Fisheries*, **23(5)**, 10–13.

Fletcher, W.J. and Sumner, N.R. (1999). Spatial distribution of sardine (*Sardinops sagax*) eggs and larvae: an application of geostatistics and resampling to survey data. *Canadian Journal of Fisheries and Aquatic Sciences*, **56(6)**, 907–914.

Ford, R.G. and Bonnell, M.L. (1997). Marine protected areas and biological distributional data: Large and small scale perspectives. *Proceedings of the Conference on California and the World Ocean*, **1**, 259.

Foucher, E., Thiam, M. and Barry, M. (1998). A GIS for the management of fisheries in West Africa: Preliminary application to the octopus stock in Senegal. *South African Journal of Marine Science*, **20**, 337–346.

Fowler, C.W., Treml, E. and Smillie, H. (1999). Georeferencing the legal framework for a web based regional ocean management geographic information system. In *Proceedings of CoastGIS 1999 GIS and New Advances in Integrated Coastal Management*, September 1999, Brest France.

Fowler, C.W. (1999). A spatial basis for managing fisheries. In *Proceedings of the 17th Lowell Wakefield Fisheries Symposium: Spatial Processes and Management of Fish Populations*, October 1999, Anchorage, Alaska.

Fox, D.S. and Starr, R.M. (1996). Comparison of commercial fishery and research catch data. *Canadian Journal of Fisheries and Aquatic Sciences*, **53**, 2681–2694.

France, A. (1999). The functions of the New Zealand Ministry of Fisheries' observer programme. In *Proceedings of the International Conference on*

*Integrated Fisheries Monitoring*. February 1999, Sydney Australia, FAO Rome, 325–328.

Franklin, F.L. (1999). Applying GIS and digital maps to real life problems in the coastal zone. *Underwater Technology*, **23(4)**, 187–189.

Freon, P., Berman, S., Kleinsmith, D., Demarcq, H., Cury, P., Roy, C., Barange, M., Shillington, F. and Penven, P. (1999). The VIBES project: combining a fishery GIS with 3D and individual-based models. In *Proceedings of the First International Symposium on Geographic Information Systems (GIS) in Fishery Science*. T. Nishida, P.J. Kailola and C.E. Hollingworth, eds. Fishery GIS Research Group, Saitama, Japan.

Fukushima, M. (1994). Spawning migration and redd construction of Sakhalin taimen, *Hucho perryi* (Salmonidae) on northern Hokkaido Island. *Japan Journal of Fish Biology*, **44(5)**, 877–888.

Gaertner, J.C., Mazouni, N., Sabatier, R. and Millet, B. (1999). Spatial structure and habitat associations of demersal assemblages in the Gulf of Lions: a multicompartmental approach. *Marine Biology*, **135(1)**, 199–208.

Gandin, L.S. (1963). Translation: Objective Analysis of Meteorological Fields. Israel Programme for Scientific Translations. Jerusalem, 1965.

Gaol, J.L. and Manurung, D. (2000). El Nino southern oscillation (ENSO) impact on sea surface temperature (SST) derived from satellite imagery and its relationship on tuna fishing ground in the South Java seawaters. *GIS At Development Magazine*. On line: http://www.gisdevelopment.net/magazine/

Garibaldi, L. and Caddy, J.F. (1998). Biogeographic characterization of Mediterranean and Black Seas faunal provinces using GIS procedures. *Ocean and Coastal Management*, **39**, 211–227.

Geist, D.R. and Dauble, D.D. (1998). Redd site selection and spawning habitat use by fall chinook salmon: the importance of geomorphic features in large rivers. *Environmental Management*, **22(5)**, 655–669.

Georgakarakos, S., Haralabus, J., Valavanis, V., Arvanitidis, C. and Koutsoubas, D. (2002). Prediction of fisheries exploitation stocks of *Loliginids* and *Ommastrephids* in Greek waters using uni and multivariate time series analysis techniques. *Bulletin of Marine Science* (in press).

Giske, J., Huse, G. and Fiksen, O. (1998). Modelling spatial dynamics of fish. *Reviews in Fish Biology and Fisheries*, **8(1)**, 57–91.

Gislason, H., Sinclair, M., Sainsbury, K. and O'Boyle, R. (2000). Symposium overview: incorporating ecosystem objectives within fisheries management. *ICES Journal of Marine Science*, **57(3)**, 468–475.

Gonzalez, A. and Marin, V.H. (1998). Distribution and life cycle of *Calanus chilensis* and *Centropages brachiatus* (Copepoda) in Chilean coastal waters: A GIS approach. *Marine Ecology Progress Series*, **165**, 109–117.

Govoni, J.J. (1993). Flux of Larval Fishes Across Frontal Boundaries: Examples from the Mississippi River Plume Front and the Western Gulf Stream Front in winter. Contributions in Science, Natural History Museum of Los Angeles County. *Bulletin of Marine Science*, **53**, 538–566.

Guillaumont, B. and Durand, C. (1999). Integration et gestion de donnees reglementaires dans un SIG: analyse appliquee au cas des cotes Francaises. In *Proceedings of CoastGIS 1999 GIS and New Advances in Integrated Coastal Management*, September 1999, Brest, France.

Guisan, A. and Zimmermann, N.E. (2000). Predictive habitat distribution models in ecology. *Ecological Modelling*, **135(2–3)**, 147–186.

Gupta, M.C., Murali, O.M., Mantry, P., Oza, S.R. and Nayak, S. (2000). Coastal zone management plan for Gujarat. *GIS At Development Magazine*, On line: http://www.gisdevelopment.net/magazine/

Habbane, M., El-Sahb, M.I. and Dubois, J.M. (1997). Determination of potential for aquaculture activities via passive teledetection and a grid based geographical information system: Application to coastal waters to the Baie des Chaleurs (Eastern Canada). *International Journal of Remote Sensing*, **18(16)**, 3439–3457.

Hassen, M.B. and Prou, J. (2001). A GIS based assessment of potential aquacultural nonpoint source loading in an Atlantic bay (France). *Ecological Applications*, **11(3)**, 800–814.

Hawks, M.M., Stanovick, J.S. and Caldwell, M.L. (2000). Demonstration of GIS capabilities for fisheries management decisions: Analysis of acquisition potential within the Meramec River Basin. *Environmental Management*, **26**, 25–34.

Hooge, P.N., Carlson, P.R. and Taggart, S.J. (2000a). The analysis of fish movement and habitat selection utilizing the animal movement ArcView extension. In *Proceedings of the Fisheries GIS Symposium of the American Fisheries Society*, August 2000, St. Louis, MO.

Hooge, P.N., Eichenlaub, W.M. and Solomon, E.K. (1999). Integrating GIS with the statistical analysis and modelling of animal movements in the marine environment. In *Proceedings of the 17th Lowell Wakefield Fisheries Symposium: Spatial Processes and Management of Fish Populations*, October 1999, Anchorage, Alaska.

Hooge, P.N., Eichenlaub, W.M. and Solomon, E.K. (2000b). Using GIS to analyze animal movements in the marine environment. http://www.absc.usgs.gov/glba/gistools/Anim_Mov_UseMe.pdf

Huettmann, F. and Diamond, A.W. (2001). Seabird colony locations and environmental determination of seabird distribution: a spatially explicit breeding seabird model for the Northwest Atlantic. *Ecological Modelling*, **141(1–3)**, 261–298.

Hyrenbach, K.D., Forney, K.A. and Dayton, P.K. (2000). Marine protected areas and ocean basin management. *Aquatic Conservation: Marine and Freshwater Ecosystems*, **10**, 437–458.

Isaak, D.J. and Bjornn, T.C. (1996). Movements of northern squawfish in the tailrace of a lower Snake River dam relative to the migration of juvenile anadromus salmonids. *Transactions of the American Fisheries Society*, **125**, 780–793.

Isaak, D.J. and Hubert, W.A. (1997). Integrating new technologies into fisheries science: the application of geographic information systems. *Fisheries*, **22(1)**, 6–10.

Kapetsky, J.M. (1989). A geographical information system for aquaculture development in Johor state. FAO Technical Cooperation Programme, FI:TCP/MAL/6754. FAO, Rome, Italy.

Kapetsky, J.M. (1994). A strategic assessment of warm water fish farming potential in Africa. CIFA Technical Paper, No. 27, p. 67.

Kapetsky, J.M. and Nath, S.S. (1997). A strategic assessment of the potential for freshwater fish farming in Latin America. COPESCAL Technical Paper No. 10, p. 128, FAO, Rome, Italy.

Kapetsky, J.M., Hill, J.M. and Worthy, L.D. (1988). A Geographical Information System for catfish farming development. *Aquaculture*, **68**, 311–320.

Kapetsky, J.M., McGregor, L. and Nanne, E.H. (1987). A geographical information system to plan for aquaculture: A FAO-UNEP/GRID study in Costa Rica. FAO Fisheries Technical Paper No. 287, p. 51, FAO, Rome, Italy.

Karp, W.A. and McEldery, H. (1999). Catch monitoring by fisheries observers in the United States and Canada. In *Proceedings of the International Conference on Integrated Fisheries Monitoring*. February 1999, Sydney Australia, FAO Rome, pp. 261–284.

Kawai, H. and Sasaki, M. (1962). On the hydrographic conditions accelerating the skipjack's northward movement across the Kuroshio front. *Bulletin of Tohoku Regional Research Institute of Fisheries Science*, **20**, 1–27.

Keleher, C.J. and Rahel, F.J. (1996). Thermal limits to salmonid distributions in the Rocky Mountain region and potential habitat loss due to global warming: a geographic information system (GIS) approach. *Transactions of the American Fisheries Society*, **125**, 1–13.

Kemp, Z. and Meaden, G. (1996). Monitoring fisheries using a GIS and GPS. In *Proceedings of the Association for Geographic Information: Geographic Information Towards the Millennium*, September 1996, Birmingham, UK, pp. 66–72.

Kennelly, S.J. (1999). The role of fisheries monitoring programmes in identifying and reducing problematic bycatches. In *Proceedings of the International Conference on Integrated Fisheries Monitoring*. February 1999, Sydney Australia, FAO, Rome, pp. 75–82.

Kieser, R., Langford, G. and Cooke, K. (1995). The use of Geographic Information Systems in the acquisition and analysis of fisheries acoustic data. In *Proceedings of the ICES International Symposium on Fisheries and Plankton Acoustics*. June 1995, Aberdeen Scotland.

Kiyofuji, H., Saitoh, S. and Sakurai, Y. (1998). A visualisation of the variability of spawning ground distribution of Japanese common squid (*Todarades pacificus*) using marine GIS and satellite data sets. *International Archives of Photogrammetry and Remote Sensing*, **32**, 882–886.

Koutsoubas, D., Valavanis, V. and Georgakarakos, S. (1998). Study of cephalopod resources dynamics in Eastern Mediterranean using GIS. In *Proceedings of GISPlaNET 98: International Conference and Exhibition on Geographic Information*, September 1998, Lisbon, Portugal.

Kracker, L.M. (1999). The geography of fish: the use of remote sensing and spatial analysis tools in fisheries research. *The Professional Geographer*, **51(3)**, 440–450.

Kraus, G., Moller, A., Trella, K. and Kester, F.W. (2000). Fecundity of Baltic cod: temporal and spatial variation. *Journal of Fish Biology*, **56(6)**, 1327–1341.

Kourti, N., Shepherd, I., Schwartz, G. and Pavlakis, P. (2001). Integrating spaceborne SAR imagery in to operational systems for fisheries monitoring. *Canadian Journal of Remote Sensing (Special Issue on Ship Detection in Coastal Waters)*, **27(4)**, 291–305.

Laidre, K.L., Jameson, R.J. and DeMaster, D.P. (2001). An estimation of carrying capacity for sea otters along the California coast. *Marine Mammal Science*, **17(2)**, 294–309.

Lee, P.F., Chen, I.C. and Tseng, W.N. (1999). Distribution patterns of three dominant tuna species in the Indian Ocean. In *Proceedings of 1999 ESRI User Conference*. http://data.esri.com/library/userconf/proc99/proceed/papers/pap564/p564.htm

Lee, R.J. and Glover, R.J.O. (1998). Evaluation of the impact of different sewage treatment processes on shellfishery pollution using a geographic information system (GIS). *Water Science and Technology*, **38(12)**, 15–22.

Lee, T.N., Clark, M.E., Williams, E., Szmant, A.F. and Berger, T. (1994). Evolution of the Tortugas Gyre and its influence on recruitment in the Florida Keys. *Bulletin of Marine Science*, **54(3)**, 621–646.

Leming, T.D., May, N. and Jones, P. (1999). A Geographic Information System for near real time use of remote sensing in fisheries management in the Gulf of Mexico. NOAA National Environmental Satellite, Data, and Information Service. Final Report to US Department of Commerce. http://sgiot2.wwb.noaa.gov/COASTWATCH/reference.htm

Lluch-Belda, D., Lluch-Cota, D.B., Hernandez-Vazquez, S., Salinas-Zavala, C. and Schwartzlose, R.A. (1991). Sardine and anchovy spawning as related to temperature and upwelling in the California current system. *CalCOFI Report*, **32**, 105–111.

Logerwell, E.A. and Smith, P.E. (1999). GIS mapping of survivor's habitat of pelagic fish off California. In *Proceedings of the 17th Lowell Wakefield Fisheries Symposium: Spatial Processes and Management of Fish Populations*, October 1999, Anchorage, Alaska.

Loneragan, N.R., Kenyon, R.A., Staples, D.J., Poiner, I.R., Conacher, C.A. (1998). The influence of seagrass type on the distribution and abundance of postlarval and juvenile tiger prawns (*Penaeus esculentus* and *P. semisulcatus*) in the western Gulf of Carpentaria, Australia. *Journal of Experimental Marine Biology and Ecology*, **228(2)**, 175–195.

Longhurst, A.R. (1998). *Ecological Geography of the Sea*, p. 398, Academic Press, San Diego.

Lowry, L.F., Frost, K.J., Davis, R., DeMaster, D.P. and Suydam, R.S. (1998). Movements and behavior of satellite tagged spotted seals (*Phoca largha*) in the Bering and Chukchi Seas. *Polar Biology*, **19**, 221–230.

Macomber, M.F. (1999). Characterisation of English sole habitat at a regional scale: A GIS study of the spatial distribution of a groundfish species using logbook data from the Oregon trawl fishery. In *Proceedings of the 17th Lowell Wakefield Fisheries Symposium: Spatial Processes and Management of Fish Populations*, October 1999, Anchorage, Alaska.

Mathews, C. (1999). Interactions between frontal systems, upwelling, El Nino and the small pelagic fisheries of the Bali Straits, Lombok Straits, Alas Straits and Sepi Straits, Indonesia. In *Proceedings of the First International Symposium on Geographic Information Systems (GIS) in Fishery Science*. T. Nishida, P.J. Kailola and C.E. Hollingworth, eds. Fishery GIS Research Group, Saitama, Japan.

Matthews, P. (1999). From logbooks to laptops: the benefits that arose from the management crisis in the Canadian East Coast, offshore scallop fishery and the introduction of monitoring to the fleet. In *Proceedings of the International*

*Conference on Integrated Fisheries Monitoring.* February 1999, Sydney Australia, FAO Rome, pp. 285–290.

Maury, O. and Gascuel, D. (1999). SHADYS (simulateur halieutiqu de dynamiques spatiales), a GIS based numerical model of fisheries: Example application: the study of a marine protected area. *Aquatic Living Resources*, **12(2)**, 77–88.

McConnaughey, R.A. and Smith, K.R. (2000). Associations between flatfish abundance and surficial sediments in the eastern Bering Sea. *Canadian Journal of Fisheries and Aquatic Sciences*, **57**, 2410–2419.

McGowan, E.M., Nealon, J. and Brown, C. (1995). Needs assessment: Aquaculture GIS in northeastern Massachusetts. In *Proceedings of GIS/LIS '95, Annual Conference and Exposition*, pp. 723–729.

Meaden, G.J. (1987). Where should trout farms be in Britain? *Fish Farmer*, **10(2)**, 33–35.

Meaden, G.J. (1996a). Monitoring fisheries effort and catch using a Geographical Information System and a Global Positioning System. In *Proceedings of the 2nd World Fisheries Conference*, July 1996, Brisbane, Australia.

Meaden, G.J. (1996b). Potential for Geographical Information Systems (GIS) in fisheries management. In *Computers in Fisheries Research*. B.A. Megrey and E. Moksness, eds. pp. 41–79. Chapman and Hall, London.

Meaden, G.J. (2000). GIS in Fisheries Management. *GeoCoast*, **1(1)**, 82–101.

Meaden, G.J. and Kapetsky, J.M. (1991). Geographic Information Systems and Remote Sensing in inland fisheries and aquaculture. FAO Fisheries Technical Paper No. 318, p. 262.

Meaden, G.J. and Kemp, Z. (1997a). The Management of Commercial Fisheries with the Aid of a Computerised Catch Logging System. In C.S. Green, ed. *EEZ Technology: The Review of Advanced Technologies for the Management of EEZs Worldwide*, pp. 181–186. ICG Publishing Ltd, London.

Meaden, G.J. and Kemp, Z. (1997b). Monitoring fisheries effort and catch using a Geographical Information System and a Global Positioning System. In D.A. Hancock, D.C. Smith, A. Grant and J.P. Beumer, eds. *Developing and Sustaining World Fisheries Resources: The State of Science and Management*, pp. 238–244. CSIRO, Collingwood, Victoria, Australia.

Mehic, N., Ross, T., Al A'ali, M. and Bakiri, G. (1996). A GIS for Bahrain Fisheries Management. In *Proceedings of 1996 ESRI International User Conference*. http://www.esri.com/library/userconf/proc96/TO250/PAP240/P240.HTM

Mejias, A. (1999). Vessel monitoring system sensor applications in the Gulf of Mexico shrimp fishery. In *Proceedings of the International Conference on Integrated Fisheries Monitoring*. February 1999, Sydney Australia, FAO, Rome, pp. 291–302.

Mikol, R.M. (1999). GIS as a bycatch reduction tool. In *Proceedings of the 17th Lowell Wakefield Fisheries Symposium: Spatial Processes and Management of Fish Populations*, October 1999, Anchorage, Alaska.

Milner, N.J., Wyatt, R.J. and Broad, K. (1998). HABSCORE: Applications and future developments of related habitat models. *Aquatic Conservation: Marine and Freshwater Ecosystems*, **8(4)**, 633–644.

Mokrin, N.M., Novikov, Y.V. and Zuenko, Y.I. (1999). Seasonal distribution of the squid (*Todarodes pacificus*) according to water structure in the Japan Sea. In *Proceedings of the 17th Lowell Wakefield Fisheries Symposium: Spatial Processes and Management of Fish Populations*, October 1999, Anchorage, Alaska.

Molenaar, E.J. and Tsamenyi, M. (2000). Satellite based vessel monitoring systems for fisheries management: international legal aspects. *The International Journal of Marine and Coastal Law*, **15(1)**, 65–110.

Moses, E. and Finn, J.T. (1997). Using geographic information systems to predict North Atlantic right whale (*Eubalaena glacialis*) habitat. *Journal of Northwest Atlantic Fishery Science*, **22**, 37–46.

Murawski, S.A., Brown, R., Lai, H.L., Rago, P.J. and Hendrickson, L. (2000). Large scale closed areas as a fishery management tool in temperate marine systems: the Georges Bank experience. *Bulletin of Marine Science*, **66(3)**, 775–798.

Murray, K.T., Read, A.J. and Solow, A.R. (2000). The use of time/area closures to reduce bycatches of harbour porpoises: lessons from the Gulf of Maine sink gillnet fishery. *Journal of Cetacean Research and Manage*ment, **2(2)**, 135–141.

Mumby, P.J., Green, E.P., Edwards, A.J. and Clark, C.D. (1999). The cost effectiveness of remote sensing for tropical coastal resources assessment and management. *Journal of Environmental Management*, **55(3)**, 157–166.

Nakata, H., Kimura, S., Okazaki, Y. and Kasai, A. (2000). Implications of mesoscale eddies caused by frontal disturbances of the Kuroshio Current for anchovy recruitment. *ICES Journal of Marine Science*, **57(1)**, 143–152.

Nath, S.S., Bolte, J.P., Ross, L.G. and Aguilar-Manjarrez, J. (2000). Applications of geographical information systems (GIS) for spatial decision support in aquaculture. *Aquacultural Engineering*, **23(1–3)**, 233–278.

Nishida, T., Kailola, P.J. and Hollingworth, C.E. (2001). *Proceedings of the First International Symposium on Geographic Information Systems (GIS) in Fishery*

*Science*, Seattle, Washington, USA, March 1999. Fishery GIS Research Group, Saitama, Japan.

Nolan, C.P. (1999). *Proceedings of the International Conference on Integrated Fisheries Monitoring*, February 1999, Sydney Australia, FAO, Rome, p. 378, M43.

O'Brien-White, S. and Thomason, C.S. (1999). Evaluating fish habitat in a South Carolina watershed using GIS. *NCASI Technical Bulletin*, **2(781)**, 391.

Olson, C.M. and Orr, B. (1999). Combining tree growth, fish and wildlife habitat, mass wasting, sedimentation, and hydrologic models in decision analysis and long term forest land planning. *Forest Ecology and Management*, **114(2)**, 339–348.

Palmer, H. (1998). GIS applications to maritime boundary definitions: international diplomacy on and under the sea. In *Proceedings of 1998 ESRI International User Conference*, San Diego CA.

Parsons, L.S. (1993). Management of marine fisheries in Canada. *Canadian Bulletin of Fisheries and Aquatic Sciences*, **225**, 1–782.

Pauly, D., Christensen, V., Froese, R., Longhurst, A., Platt, T., Sathyendranath, S., Sherman, K. and Watson, R. (2000). Mapping fisheries onto marine ecosystems: a proposal for a consensus approach for regional, oceanic and global integrations. In *Proceedings of ICES 2000 Annual Science Conference*, September 2000, Bruges, Belgium. ICES CM 2000/T:14.

Paw, J.N., Diamante, D.A.D., Robles, N.A., Chua, T.E., Quitos, L.N. and Cargamento, A.G.A. (1992). Site selection for brackish water aquaculture development and mangrove reforestation in Lingayen Gulf, Philippines using geographic information systems. In *Proceedings of Canadian Conference on GIS*, March 1992, Ottawa, Canada.

Pereira, J.M.F. (1999). Control of the Portuguese artisanal octopus fishery. In *Proceedings of the International Conference on Integrated Fisheries Monitoring*. February 1999, Sydney Australia, FAO, Rome, pp. 369–378.

Pertierra, J.P., Recasens, L. and Valavanis, V. (2001). Development of a GIS platform for the compilation and analysis of demersal species biomass off the Catalan coast (NW Mediterranean Sea). In *Proceedings of the First International Symposium on Geographic Information Systems (GIS) in Fishery Science*. T. Nishida, P.J. Kailola, and C.E. Hollingworth, eds. pp. 134–142. Fishery GIS Research Group, Saitama, Japan.

Petitgas, P. (1993). Geostatistics for fish stock assessments: a review and an acoustic application. *ICES Journal of Marine Science*, **50**, 285–298.

Piatkowski, U., Pierce, G.J. and Morais da Cunha, M. (2001). Impact of cephalopods in the food chain and their interaction with the environment and fisheries: an overview. *Fisheries Research*, **52(1–2)**, 5–10.

Pierce, G.J., Bailey, N., Stratoudakis, Y. and Newton, A. (1998). Distribution and abundance of the fished population of *Loligo forbesi* in Scottish waters: analysis of research cruise data. *ICES Journal of Marine Science*, **55**, 14–33.

Pierce, G.J., Wang, J. and Valavanis, V. (2002). Application of GIS to cephalopod fisheries. *Bulletin of Marine Science* (in press).

Platt, T. and Sathyendranath, S. (1988). Oceanic primary production: estimation by remote sensing at local and regional scales. *Science*, **241**, 1613–1620.

Platt, T. and Sathyendranath, S. (1999). Spatial structure of pelagic ecosystem processes in the global ocean. *Ecosystems*, **2**, 384–394.

Polacheck, T. (1990). Year around closed areas as a management tool. *Natural Resource Modelling*, **4(3)**, 327–354.

Populus, J., Hastuti, W., Martin, J.L.M., Guelorget, O., Sumartono, B. and Wibowo, A. (1995). Remote sensing as a tool for diagnosis of water quality in Indonesian seas. *Ocean and Coastal Management*, **27(3)**, 197–215.

Porter, M.D. and Fisher, W.L. (2000). Directions in Fishery Geographic Information Systems. In *Proceedings of the Fisheries GIS Symposium of the American Fisheries Society*, August 2000, St. Louis, MO.

Price, A.R.G., Klaus, R., Sheppard, C.R.C., Abbiss, M.A., Kofani, M. and Webster, G. (2000). Environmental and bioeconomic characterisation of coastal and marine systems of Cameroon, including risk implications of the Chad/Cameroon pipeline project. *Aquatic Ecosystem Health and Management*, **3(1)**, 137–161.

Ramster, J. (1994). Marine resource atlases: the implications for policymakers and planners. *Underwater Technology*, **19(4)**, 12–19.

Rivoirard, J., Simmonds, J., Foote, K., Fernandes, P. and Bez, N. (2000). Geostatistics for estimating fish abundance. Blackwell Science. p. 206.

Roberts, M.J. (1998). The influence of the environment of chokka squid *Loligo vulgaris reynaudii* spawning aggregations: Steps towards a quantified model. *South African Journal of Marine Science*, **20**, 267–284.

Rogers, K.B. and Bergersen, E.P. (1996). Application of geographic information systems in fisheries: habitat use by northern pike and largemouth bass. *American Fisheries Society Symposium*, **16**, 315–323.

Ross, L.G., Mendoza, Q.M. and Beveridge, M.C.M. (1993). The application of geographical information systems to site selection for coastal aquaculture: An example based on salmonid cage culture. *Aquaculture*, **112(2–3)**, 165–178.

Rowell, T.W., Trites, L.W. and Dawe, E.G. (1985). Distribution of short finned squid (*Illex illecebrosus*) larvae and juveniles in relation to the Gulf Stream frontal zone between Florida and Cape Hatteras. *Northwest Atlantic Fisheries Organization Science Council Studies*, **9**, 77–92.

Roy, P.S. (2001). Remote sensing training and education needs in India. *GIS At Development Magazine*. On line: http://www.gisdevelopment.net/magazine/

Rubec, P.J., Christensen, J.D., Arnold, W.S., Norris, H., Steele, P. and Monaco, M.E. (1998a). GIS and modelling: coupling habitats to Florida fisheries. *Journal of Shellfish Research*, **17(5)**, 1451–1457.

Rubec, P.J., Coyne, M.S., McMichael, R.H. Jr. and Monaco, M.E. (1998b). Spatial methods being developed in Florida to determine essential fish habitat. *Fisheries*, **23(7)**, 21–25.

Rubec, P.J., White, M., Wilder, D., McMichael, R., Coyne, M., Monaco, M.E., Smith, S.G. and Ault, J.S. (2000). Spatial methods to delineate fish distibutions in Tampa Bay and Charlotte Harbor, Florida. In *Proceedings of the Fisheries GIS Symposium of the American Fisheries Society*, August 2000, St. Louis, MO.

Russ, G.R. and Alcala, A.C. (1996). Marine reserves: rates and patterns of recovery and decline of large predatory fish. *Ecological Applications*, **6(3)**, 947–961.

Rutherford, E. and Brines, S. (1999). Progress Report Development of GIS for Great Lakes Fisheries Analysis. Institute for Fisheries Research, School of Natural Resources and Environment, University of Michigan, No. 14, FY 99, 230493. On line: http://www-personal.umich.edu/~edwardr/

Saitoh, S. (1983). Changes of rings and temporal and spatial variability of resources biomass. *Marine Science*, **15**, 274–285.

Sakurai, Y., Kiyofuji, H., Saitoh, S., Goto, T. and Hiyama, Y. (2000). Changes in inferred spawning areas of *Todarodes pacificus* (Cephalopoda: Ommastrephidae) due to changing environmental conditions. *ICES Journal of Marine Science*, **57(1)**, 24–30.

Salam, M.A. and Ross, L.G. (1999). A GIS modelling for aquaculture in southwestern Bangladesh: Comparative production scenarios for brackish and freshwater shrimp and fish. In *Proceedings of the 13th Annual Conference on Geographic Information Systems*, March 1999, Vancouver, Canada, pp. 141–145.

Salam, M.A. and Ross, L.G. (2000). Optimising sites selection for development of shrimp (*Panaeus monodon*) and mud crab (*Scylla serrata*) culture in southwestern

Bangladesh. In *Proceedings of the 14th Annual Conference on Geographic Information Systems*, March 2000, Toronto, Canada.

Santos, A.M.P., Borges, M.F. and Groom, S. (2001). Sardine and horse mackerel recruitment and upwelling off Portugal. *ICES Journal of Marine Science*, **58(3)**, 589–596.

Sanz, J.L., Lobato, A.B. and Tello, O. (1999). Problems on the definition and the meaning of some attributes in a GIS for the management of administration regulated coastal areas. In *Proceedings of CoastGIS 1999 GIS and New Advances in Integrated Coastal Management*, September 1999, Brest, France.

Schneider, D.C. (1998). Applied scaling theory. In D.L. Peterson and V.T. Parker, eds. *Ecological Scale: Theory and Applications*, pp. 253–269. Columbia University Press, NY, USA.

Schneider, D.C., Bult, T., Gregory, R.S., Methven, D.A., Ings, D.W. and Gotceitas, V. (1999). Mortality, movement and body size: critical scales for Atlantic cod (*Gadus morhua*) in the northwest Atlantic. *Canadian Journal of Fisheries and Aquatic Sciences*, **56(S1)**, 180–187.

Scientific Fishery Systems, Inc. (1999). Fisheries applications of satellite imagery. http://www.alaska.net/~scifish/ResearchAndDevelopment/ResearchAndDevelop ment.html

Service, M. and Magorrian, B.H. (1997). The extent and temporal variation of disturbance to epibenthic communities in Strangford Lough, Northern Ireland. *Journal of the Marine Biological Association of the United Kingdom*, **77(4)**, 1151–1164.

Sherman, K., Alexander, L.M. and Gold, B.D. (1990). *Large marine ecosystems: patterns, processes and yields*. American Association for Advancement of Science, Washington DC, p. 242.

Sherman, K., Alexander, L.M. and Gold, B.D. (1993). *Stress, mitigation and sustainability of large marine ecosystems*. American Association for Advancement of Science, Washington DC, p. 376.

Simpson, J.J. (1994). Remote sensing in fisheries: A tool for better management in the utilization of a renewable resource. *Canadian Journal of Fisheries and Aquatic Sciences*, **51**, 747–771.

Soh, S.K., Gunderson, D.R. and Ito, D.H. (2001). The potential role of marine reserves in the management of shortraker rockfish (*Sebastes borealis*) and rougheye rockfish (*S. aleutianus*) in the Gulf of Alaska. *Fishery Bulletin*, **99(1)**, 168–179.

Soletchnik, P., Moine, O.L., Faury, N., Razet, D., Geairon, P. and Goulletquer, P. (1999). Summer mortality of the oyster in the Bay Marennes/Oleron: spatial variability of environment and biology using a geographical information system (GIS). *Aquatic Living Resources*, **12(2)**, 131–143.

Soliday, B. (2000). The unexpected role of high-resolution satellite imagery in protecting marine habitats. *Surveying and Land Information Systems*, **60(4)**, 285–286.

Stanbury, K.B. and Starr, R.M. (1999). Applications of Geographic Information Systems (GIS) to habitat assessment and marine resource management. *Oceanologica Acta*, **22**, 699–703.

Stolyarenko, D.A. (1995). Methodology of shellfish surveys based on microcomputer Geographic Information Systems. *ICES Marine Science Symposia*, **199**, 259–266.

Stoner, A.W., Manderson, J.P. and Pessutti, J.P. (2001). Spatially explicit analysis of estuarine habitat for juvenile winter flounder: combining generalized additive models and Geographic Information Systems. *Marine Ecology Progress Series*, **213**, 253–271.

Taconet, M. and Bensch, A. (1999). Towards the use of geographic information systems as a decision support tool for the management of Mediterranean fisheries. Final Report for FAO COPEMED Project GCP/REM/057/SPA. Online: http://www.ua.es/copemed/vldocs/0000028/

Tameishi, H., Shinomiya, H., Aoki, I. and Sugimoto, T. (1996). Understanding Japanese sardine migrations using acoustic and other aids. *ICES Journal of Marine Science*, **53**, 167–171.

Treml, E., Neely, R., Hamilton, S., Fowler, C. and LaVoi, T. (1999). Georeferencing the Legal and Statutory Framework for Integrated Regional Ocean Management. In *Proceedings of 1999 ESRI International User Conference*. On line: http://data.esri.com/library/userconf/proc99/proceed/papers/pap498/p498.htm

Turk, T.A. (1999). Spatial distribution and habitat preferences of weathervane scallops (*Patinopteron caurinus*) in the Gulf of Alaska. In *Proceedings of the 17th Lowell Wakefield Fisheries Symposium: Spatial Processes and Management of Fish Populations*, October 1999, Anchorage, Alaska.

Valavanis, V., Georgakarakos, S., Koutsoubas, D., Arvanitidis, C. and Haralabus, J. (2002). Development of a marine information system for Cephalopod fisheries in the Greek Seas (Eastern Mediterranean). *Bulletin of Marine Science* (in press).

Valavanis, V.D., Georgakarakos, S. and Haralabus, J. (1998). A methodology for GIS interfacing of marine data. In *Proceedings of GISPlaNET 98 International*

*Conference and Exhibition of Geographic Information*, September 1998, Lisbon, Portugal.

Van der Lingen, C.D. (1999). Comparative spawning habitats of sardine and anchovy in the Southern Benguela upwelling region. In *Proceedings of the 17th Lowell Wakefield Fisheries Symposium: Spatial Processes and Management of Fish Populations*, October 1999, Anchorage, Alaska.

Varkentin, A.I., Buslov, A.V. and Tepnin, O.B. (1999). Characteristics of spawning and distribution of walleye pollock eggs and larvae in Western Kamchatka waters. In *Proceedings of the 17th Lowell Wakefield Fisheries Symposium: Spatial Processes and Management of Fish Populations*, October 1999, Anchorage, Alaska.

Veisze, P. and Karpov, K. (1999). Geopositioning a remotely operated vehicle for marine species and habitat analysis. In *Proceedings of 1999 ESRI International User Conference*. On line: http://www.esri.com/library/userconf/proc99/proceed/papers/pap517/p517.htm

Walters *et al.* (1992). Larval drift and growth simulator. *Fisheries Oceanography*, **1(1)**, 11–19.

Waluda, C.M. and Pierce, G.J. (1998). Temporal and spatial patterns in the distribution of squid *Loligo* spp. in United Kingdom waters. *South African Journal of Marine Science*, **20**, 323–336.

Waluda, C.M., Rodhouse, P.G., Trathan, P.N. and Pierce, G.J. (2001). Remotely sensed mesoscale oceanography and the distribution of *Illex argentinus* in the South Atlantic. *Fisheries Oceanography*, **10(2)**, 207–216.

Ward, T.J., Vanderklift, M.A., Nicholls, A.O. and Kenchington, R.A. (1999). Selecting marine reserves using habitats and species assemblages as surrogates for biological diversity. *Ecological Applications*, **9(2)**, 691–698.

Webb, A.D. and Bacon, P.J. (1999). Using GIS for catchment management and freshwater salmon fisheries in Scotland: the DeeCAMP project. *Journal of Environmental Management*, **55**, 127–143.

White, D.L., Bushek, D., Porter, D.E. and Edwards, D. (1998). Geographic Information Systems (GIS) and kriging: Analysis of the spatial and temporal distributions of the oyster pathogen *Perkinsus marinus* in a developed and an undeveloped estuary. *Journal of Shellfish Research*, **17(5)**, 1473–1476.

Williamson, N. and Traynor, J. (1996). Application of a one dimensional geostatistical procedure to fisheries acoustic surveys of Alaskan pollock. *ICES Journal of Marine Science*, **53**, 423–428.

Wright, P.J., Jensen, H. and Tuck, I. (2000). The influence of sediment type on the distribution of the lesser sandeel, *Ammodytes marinus*. *Journal of Sea Research*, **44**, 243–256.

Xavier, J.C., Rodhouse, P.G., Trathan, P.N. and Wood, A.G. (1999). A Geographical Information System (GIS) Atlas of cephalopod distribution in the Southern Ocean. *Antarctic Science*, **11**, 61–62.

Zheng, X., Pierce, G.J. and Reid, D.G. (2001). Spatial patterns of whiting abundance in Scottish waters and relationships with environmental variables. *Fisheries Research*, **50**, 259–270.

# CHAPTER FOUR

# Instead of an Epilogue

After three decades of technical development we have now reached the impressive situation in which computer and space technology have made it possible for us to sample, store, analyse, simulate, visualise and integrate enormous amounts of marine spatial digital data. We have the Global Positioning System (GPS) through which obtained precise locational measurements are being used for endless applications. We have Remote Sensing (RS), which include a number of advanced satellite sensors providing important and diverse information on global ocean surface, airborne sensors for the acquisition of multiple information for local smaller areas and underwater data recording devices for sampling the ocean water column. These sensors use the light and sound spectra, and electromagnetic and microwave radiation to convert spectral reflectance signatures to interpretable images of ocean environment. We have sophisticated image processing and analysis techniques, which include a variety of image enhancement and interpretation steps, such as radiometric and geometric corrections, georeference and classification. We have sophisticated geostatistical techniques, which are applied to field survey sampling data to reveal valuable relations and interactions among data. We have sophisticated physical and ecological modelling techniques, which allow data assimilation and forecasting of oceanographic processes and ecosystem conditions. We have sophisticated multidimensional scientific visualisation systems, which offer a unique method for comprehending the 3D nature of marine environment. These technological advances gave a boost in our knowledge and understanding of marine world processes by providing vast amounts of digital data and tools to interpret them.

The links among the complexities of the marine environment, which is under a continual natural change and regeneration as well as a continuing human-induced pressure, are often seen as these small, difficult to discover and appropriately place, however greatly important pieces of a big complex puzzle. Scientists and environmental managers, realising the importance of the ever-changing links among the physical and biological components of our oceans, need to fully comprehend what exactly drives the phenomenal diversity of links among processes and life nourishment in Earth's liquid environment. For this purpose, we have a sophisticated new technology, Geographic Information Systems (GIS), which, although is based on a relatively simple concept, provides an invaluable mean for the discovery and explanation of those spatial-oriented links that characterise the marine world. It is the ability of GIS to integrate different layers of information that facilitates the identification of dynamic links among marine interactions. GIS, based on the 'onion' concept where data about an area are placed in thematic commonly referenced data layers, greatly facilitated the explanations of environmental links in terrestrial 'static' environments during the last three or four decades. However, during the last decade or more, when GIS technology has started migrating to sea, things became more complex. Multiple data storage (in context and format) became more complicated, the third dimension became necessary in spatial analysis and integrations among different data became more extensive. Now, these requirements pose several new technological challenges (including new software designs, new 3D GIS database management systems, new GIS integration routines as well as interoperability and seamless network working environments, metadata standards, etc.), forcing GIS technology to be further developed in order to be fully and appropriately applied in a highly diverse 3D field of applications.

However, the application of current GIS technology in Oceanography and Fisheries includes a variety of brilliant approaches. As we have seen, marine-related uses of GIS bring under a common technological, scientific and managerial framework a number of similar disciplines, merge expertise from a variety of marine sciences and provide integrated management products for many purposes. In marine studies, GIS contribute as data distribution, mapping, integration, visualisation and management tools in a great variety of studies and research, including coastal and submerged vegetation mapping, coastal bathymetry mapping, wetland research, flood and natural hazard research, coastal and open ocean oceanographic processes research, deep ocean bathymetry mapping, marine geomorphology and deep environments research, coastal fisheries and spawning grounds research, fish essential habitat mapping, MPA design, ocean fisheries research, fisheries monitoring system design and aquaculture and inland fisheries research.

In parallel with this multidisciplinary explosion of marine GIS applications, many national and international organisations have started to adapt GIS technology as their primary data management and dissemination platform. Today, such GIS tools are widely developed and placed in the Internet, concentrating under one base invaluable data, facilitating dissemination of data and communicating ready to use data into many managerial authorities and the general public. In addition, national and international spatial data infrastructures are continually organised and developed worldwide in order to reduce duplication of effort among agencies, improve quality and reduce costs related to geographic information, make geographic data more accessible to the public, increase the benefits of using available data and establish key partnerships within and among countries. Many countries and states are already benefit from such spatial data infrastructures, for example, in United States (http://www.fgdc.gov/), in Asia and the Pacific (Majid 1997), in Australia (Mooney and Grant 1997 and http://www.auslig.gov.au/asdi/), in Canada (http://cgdi.gc.ca/) and Europe (http://www.eurogi.org/). An excellent survey of national and regional spatial data infrastructure activities around the globe was conducted by Harlan J. Onsrud at the University of Maine (http://www.spatial.maine.edu/~onsrud/GSDI.htm)

Future uses of GIS technology in marine sciences will definitely increase. For example, in the Global Meeting of the Geological Society of America, which was held in Edinburgh, Scotland on June 2001 with themes on 'Earth System

Processes and Geological Research on Ocean Margins: Trends and Possibilities' (http://www.geosociety.org/meetings/edinburgh/), existing trends and an envisage of future promise identified the importance of future work towards explanation of the explicit interaction of the biosphere with the geosphere. Mapping of deep ocean environments and processes will become a first priority field of study for the explanation of such interactions. Also, the general recognition of the complex spatial nature of problems related to fisheries resources and management (e.g. overexploitation, destruction of habitats, allocation of fishing effort) will eventually result in management schemes that will integrate indicators such as species interaction with the environment, identification and protection of fish essential habitats, identification of alternative fishing grounds, etc. With a future migration from 2D to 3D integration in GIS, the amounts of data to process will grow substantially and associated techniques will be greatly altered but our modelling abilities will be extensively facilitated. As extensive knowledge on species life history cycles will be integrated with environmental processes in three dimensions, we will develop highly accurate models and computer representations of our valuable marine resources. In this effort, concentration on focal species with indicator properties of ecosystem health will greatly advance synoptic views of marine resources (Zacharias and Roff 2001). Meaden (2000) identified the future GIS uses in fisheries management, which include the electronic management of production data, the mapping of EFHs and design of marine reserves, spatial analysis prior to stock enhancement efforts, the assessment of fishing gear damage and effects of fish removal to marine ecosystems, establishment of optimum locations for deep-sea mariculture, design of flexible spatial sampling units and practices for optimum implementation of GIS in fisheries management.

During the last few years, the marine and fisheries GIS field has enriched with several excellent books while a new book is in press. '*Marine and Coastal Geographical Information Systems*' (Wright and Bartlett 2000, published by Taylor and Francis) being the first multi-author book in the field, analyses a variety of marine GIS issues and presents many excellent marine and coastal GIS applications. The brand new multi-author '*Undersea with GIS*' (Wright 2001, published by ESRI Press) includes a variety of topics on oceanographic GIS applications and is accompanied by an excellent CDROM with many valuable GIS tools, animations, datasets and more. Currently in press, the multi-author '*Geographic Information Systems in Fisheries*' (Fisher and Rahel 2002, to be published late in 2002 by the American Fisheries Society) will be an invaluable contribution to inland, aquaculture and marine fisheries GIS application fields. The time also has come for a new marine and fisheries GIS journal in order to include the excellent and rapidly developing applications in these fields, which consist the backbone of sea-related use of GIS and have many to offer for information-based management of our coastal and ocean resources. In addition, now is the time for an organised marine GIS textbook, which will be an invaluable educational and training source for the use of latest technological developments in marine resource management.

A relatively recent adopted approach in GIS modelling is the introduction of cellular automata (CA). A cellular automaton is a dynamical system in which space, time and the states of the system are all represented discretely. Space is represented by a regular lattice and each cell in the lattice can be in one of a finite number of states represented by integer number values. Each cell looks to its neighbouring cells to see which states they are in. From this, each cell uses a set of simple rules to determine the state that it should be in. Related methodologies are based on rules defining the state of each cell. These rules are stochastic, being all probabilities defined through the use of deterministic mathematical models. The theory was first introduced by John Von Neumann and Stanislaw Ulam (US Los Alamos National Laboratory) in the 1940s and gained considerable popularity in the 1960s through the work of John Conway in the game of life (Gardner 1970). In terms of structure, CA computational scheme is similar to the ones employed in the numerical manipulation of partial differential equations. The difference is that the state variable at each cell of the lattice is only allowed to assume a small set of values (typically two states per cell) and that the transition functions do not assume an algebraic form but may be deterministic or stochastic. CA models are ideally applied for diffusion modelling, considering a small number of state variables and relatively simple transition rules. The use of CA/GIS in marine studies is already introduced by several authors. For example, Bonfatti *et al.* (1994) developed a GIS embedded CA model for modelling the tidal wave in the Venice lagoon (Italy), Engelen *et al.* (1995) used CA for the integrated modelling of social/environmental systems and Voinov *et al.* (1999) developed a similar system for the integrated ecological/economic modelling of watersheds. Future use of CA in marine GIS will definitely expand since CA is an approach with growing importance in discrete dynamic systems.

Water, the essential element of life in all of its forms, the presence of which is the reason of a high diversity of species aggregation around and in ponds, lakes, rivers and the sea, seems to extend its attractiveness to GIS technology, as well! It is not strange that even among those excellent GIS applications in forestry and wildlife in terrestrial ecosystems, it is forest hydrology that attracts a good number of applications. Water sustains life but also gives new biological birth and evolution in endless forms. Now, 'liquid' applications of GIS require and force GIS technology to be further developed and expanded. It would not be too much to say that in both application (GISystems) and discipline levels (GIScience), to make the most out of GIS, mix it with water!

## 4.1 REFERENCES

Bonfatti, F., Gadda, G. and Monari, P.D. (1994). Cellular automata for modelling lagoon dynamics. In *Proceedings of EGIS/MARI 1994, 5th European Conference and Exhibition on Geographical Information Systems*, Paris France.

Earle, S. and Wright, D. (2001). *Undersea with GIS*. ESRI Press, California, USA.

Engelen, G., White, R., Uljee, I. and Drazan, P. (1995). Using cellular automata for integrated modelling of socioenvironmental systems. *Environmental Monitoring and Assessment*, **30**, 203–214.

Fisher, W.L. and Rahel, F.J. (2002). *Geographic Information Systems in Fisheries*. American Fisheries Society, in press.

Gardner, M. (1970). The fantastic combinations of John Conway's new solitaire game 'life'. *Scientific American*, **223**, 120–123.

Majid, D.A. (1997). Geographical data infrastructure in Asia and the Pacific. In D. Rhind, ed. *Framework for the World*, p. 206–210. GeoInformation International, Cambridge.

Meaden, G.J. (2000). GIS in Fisheries Management. *GeoCoast*, **1(1)**, 82–101. On line: http://www.theukcoastalzone.com/geocoast/

Mooney, D.J. and Grant, D.M. (1997). The Australian National Spatial Data Infrastructure. In D. Rhind, eds. *Framework for the World*. pp. 197–205. GeoInformation International, Cambridge.

Voinov, A.A., Costanza, R., Wainger, L.A., Boumans, R.M.J., Villa, F., Maxwell, T. and Voinov, H. (1999). Integrated ecological economic modelling of watersheds. *Environmental Modelling and Software*, **14**, 473–491.

Wright, D. (2001). *Undersea with GIS*. ESRI Press, California, USA.

Wright, D. and Bartlett, D. (2000). *Marine and Coastal Geographical Information Systems*. Taylor and Francis, Philadelphia, USA, p. 320.

Zacharias, M.A. and Roff, C. (2001). Use of focal species in marine conservation and management: A review and critique. *Aquatic Conservation: Marine and Freshwater Ecosystems*, **11**, 59–76.

# GIS Routines for Chapter Two

Annex I contains macro routines in Arc Macro Language (AML) for use with any version of ESRI's ARC/INFO GIS package. AMLs will need various modifications depending on the geographical area, projection files and image formats involved as well as local directory structures, integration tolerances, etc. These routines were developed by the author for use in GIS developments in European and Greek National projects and they are freely distributed to provide ideas of GIS programming for the study of oceanographic processes and bottom substrate mapping and to be used as a base for further developments. Please, feel free to contact the author by e-mail for any questions on these routines. Basic experience with AML and MENU development within ARC/INFO is assumed.

The following three files are used for AVHRR SST image downloading and processing. The system uses DLR's GISIS network interface to DLR's image archives:

```
-------------------------------
/* sstupdate.aml
&echo &off
&term 9999
w /avhrr
&thread &create ~
&m /avhrr/sstupdate &size 800 250 &stripe 'SST DATABASE UPDATE SYTEM' &position &cc
&thread &delete &self
-------------------------------


-------------------------------
7 sstupdate.menu

 %draw1  %draw2  %draw3

 %exit

%draw1 button keep 'View latest SST in database' &s img := [getgrid /avhrr/images/00week 'Select
SST grid:'];ap;display 9999;mape %img%;shadeset colornames;gridshades %img% # sst.rmp;move .1
.1;text %img%;&m /avhrr/images/basta &size 160 100 &stripe 'MAIN MENU'
```

%draw2 button keep 'Start GISIS' w /avhrr/images/00week;/common1/gisis/gisis
%draw3 button keep 'Start image processing & View processed grid' w /avhrr/images/00week;&s img1
:= [getimage /avhrr/images/00week/ -ERDAS 'Select SST image:'];imagegrid %img1% new99;&r
procsst new99;&m /avhrr/images/basta &size 160 100 &stripe 'MAIN MENU'
%exit button keep icon /avhrr/upsst.icon Draw q

------------------------------


------------------------------

```
/*procsst.aml
&args grd
   grid
   display 9999
   mape %grd%
   shadeset colornames
   gridshades %grd% # sst.rmp
   setwindow 2231 522 3107 1264
   %grd%gr = %grd%
   q
&if [exists %grd% -grid] &then
 &do
   adjust %grd%gr test.lnk gr%grd%
   kill %grd%gr all
   project grid gr%grd% %grd%gr prj.prj
   kill gr%grd% all
   kill %grd% all
 &end
   grid
   mape %grd%gr
   shadeset colornames
   gridshades %grd%gr # sst.rmp
   &messages &pop
   &s .repname = [response 'Give a name to the new grid:']
   &messages &on
   rename %grd%gr %.repname%
```

------------------------------

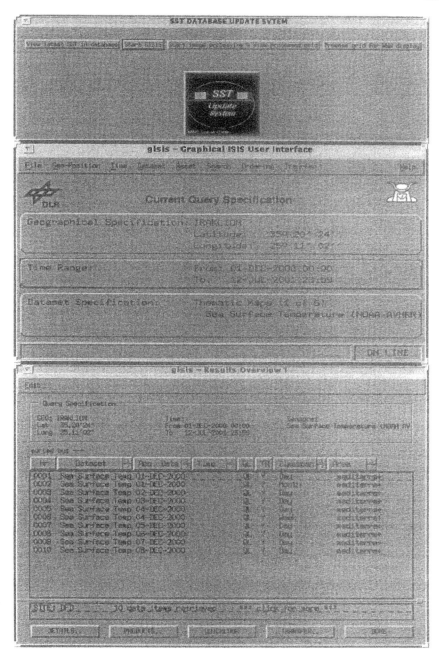

**Figure I.1.** GIS-based AVHRR SST image processing system (introduction to GIS database) using DLR's GISIS network interface to DLR's SST image archives. The system provides automated image downloading, georeference and grid conversion.

The following three files are used for SeaWiFS image downloading and processing. The system uses NASA's SeaWiFS Project website and image archives:

```
------------------------------
/* chlupdate.aml
&echo &off
&term 9999
w /seawifs
&thread &create ~
&m /seawifs/chlupdate &size 800 250 &stripe 'CHL DATABASE UPDATE SYTEM' &position &cc
&thread &delete &self
------------------------------


------------------------------
7 chlupdate.menu

  %draw1  %draw2  %draw3

  %exit

%draw1 button keep 'View latest CHL in database' &s img := [getgrid /seawifs/images/00week 'Select
CHL grid:'];ap;display 9999;mape %img%;shadeset colornames;gridshades %img% # chl.rmp;move .1
.1;text %img%;&m basta &size 160 100
%draw2 button keep 'Connect to SeaWiFS Server' netscape 'http://seawifs.gsfc.nasa.gov/cgibrs/
level3.pl?TYP=chl'
%draw3 button keep 'Start image processing & View processed grid' w /seawifs/images/00week;&s
img1 := [getimage /seawifs/images/00week/ -ERDAS 'Select SST image:'];imagegrid %img1%
new99;&r procsst new99;&m /seawifs/images/basta &size 160 100 &stripe 'MAIN MENU'
%exit  button keep icon /seawifs/upchl.icon Draw q
------------------------------


------------------------------
/* procchl.aml
&args grd
  grid
  display 9999
  mape %grd%
  shadeset colornames
  gridshades %grd% # clas.rmp
  setwindow 2263 1409 2383 1497
  %grd%1 = %grd%
  q
  gridwarp %grd%1 link.lnk %grd%2
&if [exists %grd% -grid] &then
 &do
   project grid %grd%2 %grd%3 prj.prj
   latticeclip %grd%3 /covers/fishareas %grd%4 extent
 &end
```

```
&else
 &do
  &ty something goes wrong...
 &end
  grid
  %grd%gr = con(isnull(%grd%4),%grd%3)
  kill (! %grd% %grd%1 %grd%2 %grd%3 %grd%4 !) all
  mape %grd%gr
  shadeset colornames
  gridshades %grd%gr # clas.rmp
  &messages &pop
  &s .repname = [response 'Give a name to the new grid:']
  &messages &on
  rename %grd%gr %.repname%
```

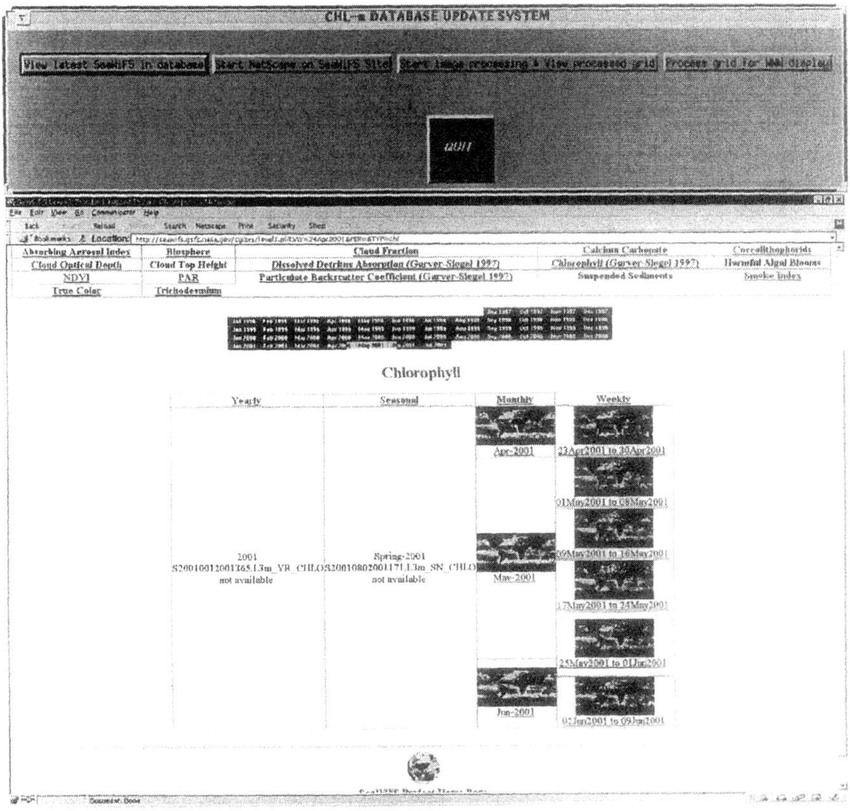

-------------------------------

**Figure I.2.** GIS-based SeaWiFS CHL image processing system (introduction to GIS database) using NASA's GSFC SeaWiFS website with image archives. The system provides automated image downloading, georeference and grid conversion.

The following five files are used for manipulation of AVHRR and/or SeaWiFS images in relation to the upwelling process. The fifth file (upwell_cla.aml) uses a predefined classification table (LUT) for image manipulation. These files consist the main core of the Upwelling Identification and Measurement System (UIMS), a GIS application developed for the study of upwellings and gyre formations in SE Mediterranean:

```
------------------------------
/* upwell.aml
&echo &off
&term 9999
grid
display 9999 3
&thread &create &m upwell &stripe 'U.I.M.S.' &size &canvas 330 450 ~
&pos &ur &display &ur
&thread &delete &self
------------------------------
```

```
------------------------------
7 upwell.menu

%anim
    ...and note images showing upwellings
%selec
%are
%locale
%mea
%wind
%qqq

/*definitions

%anim button keep 'View A/I animation' &r pseudo /*an AML drawing SST grids/*
%selec button keep 'Select and View Upwelling Image' &s .grid := [getgrid * 'Select a grid to view:'];
mape %.grid%; clear; shadeset colornames; gridshades %.grid% # sst.rmp; &m zoom
%are button keep 'Measure Upwelling Area' &messages &popup; mapunits meters; measure area
%locale button keep 'Find Upwelling Location' &messages &popup; mapunits meters; measure where
%mea button keep 'Calculate Upwelling Mean Temperature' &messages &on; &r testa.aml
%wind button keep 'Calculate Wind Sector Mean Temperature' &r testb.aml
%qqq button keep EXIT q; q
------------------------------
```

```
------------------------------
/* testa.aml
&type 'Using the mouse, select upwelling area by drawing a polygon...'
testa = selectpolygon(%.grid%, *, inside)
describe testa
------------------------------
```

```
------------------------------
/* testb.aml
&s .cover := [getcover '/common/vasilis/rs/wind*' -poly 'Select a wind sector for analysis:']
setwindow %.cover%
outgrd1 = %.grid%
outgrd2 = con(isnull(testa),outgrd1)
outgrd = select(outgrd2, 'value > 0 & value < 255')
describe outgrd
kill (!testa outgrd1 outgrd2 outgrd!) all
------------------------------


------------------------------
/* upwell_cla.aml
/*AML that uses a LUT for image classification based on predetermined values
/* e.g. every .5 degrees Celcius
/*runs in Arc:
grid
display 9999 3
&s .grid := [getgrid * 'Select a grid to process:']
mape %.grid%
clear
shadeset colornames
gridshades %.grid% # sst.rmp
arc latticepoly %.grid% grdpol range ../../sstcode.lut
linecolor 0
arcs grdpol
mape *
clear
gridshades %.grid% # sst.rmp
arcs grdpol
resel grdpol poly many *
writesel selec.sel grdpol poly
arc reselect grdpol grdpol1 poly selec.sel poly
clear
gridshades %.grid% # sst.rmp
arcs grdpol1
arc latticeclip %.grid% grdpol1 octlat extent
setwindow *
outgrd1 = con(isnull(octlat), %.grid%)
outgrd = select(outgrd1, 'value > 0 & value < 255')
&describe octlat
&type '**********'
&type 'Mean of upwelling...'
&lv %grd$mean%
&type '**********'
&type 'Area of upwelling...'
list grdpol1.pat
&type '**********'
```

&describe outgrd
&type 'Mean of upwelling surrounding area...'
&lv %grd$mean%
mapunits meters
&type '***********'
&type 'Location of upwelling...'
measure where
&type '***********'
kill (!grdpol grdpol1 octlat outgrd1 outgrd!) all
&type 'Done...'
-----------------------------

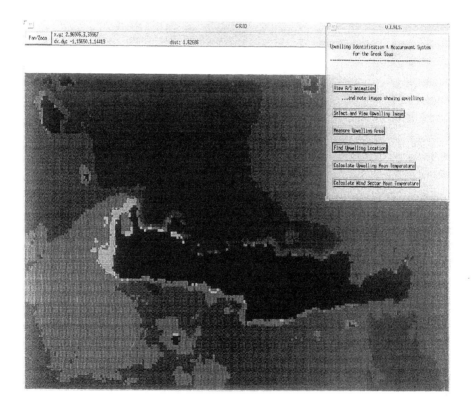

**Figure I.3.** Upwelling identification and measurement through GIS. The system uses manual and automated methods for selection of upwelling from AVHRR SST images, measurement of upwelling extent, location and temperature difference inside and outside of the upwelling area.

The following three files are used for the manipulation of RoxAnn® sonar sediment data and aerial photography interpretation for submerged vegetation mapping. These files describe the introduction of sonar data to point coverages from comma separated value ASCII files (*.csv) and interpolation of values in GIS. They also describe a georeference technique for aerial photos and extraction of polygon habitat types using on screen digitising.

------------------------------

```
/* creates a point cover with RoxAnn data
/* stored on an ASCII csv files
/* runs in Arc:
create rox grdd
generate rox
input rox.csv
points
q
build rox points
tables
define roxx.dat
ROX-ID 4 5 B
LON 8 10 F 8
LAT 8 10 F 8
CODE 1 1 I
DEP 8 8 F 2
~
add ROX-ID LON LAT CODE DEP from rox.csv
joinitem rox.pat roxx.dat rox.pat ROX-ID
project cover rox rox1 dd_tr.prj
```
------------------------------


------------------------------
```
/* topog.aml
/* interpolation using topogrid
/* runs in Arc:
topogrid sedgrd 15
enforce off
datatype spot
point POIN code
contour AKTI code
boundary SITE
end
~
topogrid depgrd 15
datatype spot
enforce off
point POIN dep
contour AKTI dep
```

boundary SITE
end
~
latticecontour depgrd depcov 5
------------------------------


------------------------------
/* digit.aml
/* registers and temporarily projects an aerial photo
/* for on-screen digitizing of polygons
/* runs in Arc:
register 5652.jpg grtr 1 composite 1 2 3
create fotocov grtr
build fotocov lines
ae;display 9999 4
ec fotocov
mape image 5652.jpg
ef arcs
de arcs
mapwarp 5652.jpg
image 5652.jpg composite 1 2 3
PAGEUNITS CM
MAPUNITS METERS
mapscale 5000
draw
arcsnap off
nodesnap closest 50
add
save
build nodups nodiffs
save
ef poly
additem fotocode 1 1 I
sel all
sel many *
calc fotocode = 1
sel all
sel many *
calc fotocode = 2
sel all
save
q
------------------------------

# GIS Routines for Chapter Three

Annex II contains macro routines in AML for use with any version of ESRI's ARC/INFO GIS package. AMLs will need various modifications depending on the geographical area, projection files and image formats involved as well as local directory structures, integration tolerances, etc. These routines were developed by the author for use in GIS developments in European and Greek National projects and they are freely distributed to provide ideas of GIS programming for the study of fish population dynamics and to be used as a base for further developments. Please, feel free to contact the author by e-mail for any questions on these routines. Basic experience with AML and MENU development within ARC/INFO is assumed.

The following file (spawn.aml) creates grids of temperature (SST), salinity (SSS) and substrate according to species preferred spawning conditions, converts these grids to polygon coverages and integrates the resulted coverages to obtain species spawning locations. Species preferred spawning conditions include certain ranges in SST, SSS, substrate type and bathymetry.

```
--------------------------------
/*spawn.aml
/*runs in Arc:

grid
outgrdt = select(jansst, 'value > 123 and value < 220')
outgrds = select(jansss, 'value > 89 and value < 153')
outsed = select(sedgrd, 'value > 1 and value < 3')
q

gridpoly outgrdt outcovt
gridpoly outgrds outcovs
gridpoly outsed outcov
intersect outcovt outcovs outcovts
intersect outcov outcovts spawn1
intersect spawn1 bathy50 spawn

kill (!outgrdt outgrds outsed outcovt outcovs outcov outcovts!) all

ap
```

```
display 9999
mape spawn
arcs gr
polygonshades spawn 4
```
------------------------------


The following file (migrat.aml) applies cost allocation functions in a SST image in order to output species preferred migration corridors through SST values. Two points must be specified by the user: (1) a start point (e.g. spawning grounds); and (2) an end point (e.g. major fishing areas). The AML can be applied to a combination of images (e.g. SSS, Chl-a, currents, etc.) or to a time series of images, with constraints taken from species life history data.


------------------------------
```
/*migrat.aml
/*runs in GRID:

display 9999
arc w /weekly/rs/images/
&s .grid := [getgrid * 'Select weekly SST image:']
mape %.grid%
shadeset colornames
gridshades %.grid% # sst.rmp
mape *
clear
gridshades %.grid% # sst.rmp
setwindow *

&describe %.grid%
%.grid%1 = %grd$zmax% - %.grid%

&ty Get start point

&getpoint &current &map
&set .costpath$x = %pnt$x%
&set .costpath$y = %pnt$y%
&set .costpath$startx = %.costpath$x%
&set .costpath$starty = %.costpath$y%

&ty Get end point

&getpoint &current &map
&set .costpath$x = %pnt$x%
&set .costpath$y = %pnt$y%
&set .costpath$endx = %.costpath$x%
&set .costpath$endy = %.costpath$y%
```

```
&set x1 = %.costpath$startx%
&set y1 = %.costpath$starty%
&set x2 = %.costpath$endx%
&set y2 = %.costpath$endy%

xxcdist = costdistance(selectpoint(%.grid%1,%x1%,%y1%),%.grid%1,xxcback)

xxpath = con(costpath(selectpoint(%.grid%1,%x2%,%y2%),xxcdist,xxcback),2)

xxtotalc = int(%.grid%1)
xxallo1 = costallocation(selectpoint(xxtotalc,%x1%,%y1%),xxtotalc,xxaccum1)
xxallo2 = costallocation(selectpoint(xxtotalc,%x2%,%y2%),xxtotalc,xxaccum2)

xxcorr = con(select(slice(corridor(xxaccum1,xxaccum2),eqinterval,150), 'value eq 1'),2)

xxdispopt = gridline(xxpath)
xxdispcor = gridpoly(xxcorr)

resel xxdispcor poly grid-code = 2
polygonshades xxdispcor 33
asel xxdispcor poly
linesymbol 2
polygons xxdispcor
arclines xxdispopt 5

markersymbol 1
marker %.costpath$startx% %.costpath$starty%
markersymbol 2
marker %.costpath$endx% %.costpath$endy%

&do j &list xxtotalc xxallo1 xxallo2 xxaccum1 xxaccum2 xxcback xxcdist
 &if [exists %j% -grid] &then
   arc kill %j% all
&end

kill (!xxcorr xxpath xxdispcor %.grid%1!) all
list xxdispopt.aat length
kill xxdispopt all
-------------------------------
```

The following file (efh.aml) uses a series of selections, conversions and integrations among vector and raster datasets for the identification of EFH (see also Table 3.3). The idea is to identify species major catch areas, extract environmental conditions in these areas (minimum and maximum values), select these value ranges from satellite images, convert the resulted areas to polygons, and finally integrate all polygons in order to identify the common areas that satisfy all species preferred environmental conditions.

```
--------------------------------
/*efh.aml
/*runs in Arc:
/*uses a polygon coverage for fish catch data (catch),
/*a polygon coverage for fleet fishing activity areas (fleet),
/*satellite images of SST(sstgrd) and Chl-a (chlgrd), and SSS(sssgrd),
/*and a polygon coverage for bathymetric contours (bathym)

ap
resel catch poly species > 0
writeselect catch.sel catch poly
resel bathym poly contour = 100
writeselect bathym.sel bathym poly
q
reselect catch catchsp poly catch.sel poly
reselect bathym bathsp poly bathym.sel poly

intersect catchsp bathsp catsbath # 0.00000001
intersect catsbath fleet fishareas # 0.00000001

grid
sstout = selectpolygon(sstgrd, fishareas)
&describe sstout
sstout1 = select(sstgrd, 'value > %grd$zmin% and value < %grd$zmax%')
sstcov = gridpoly(sstout1)

sssout = selectpolygon(sssgrd, fishareas)
&describe sssout
sssout1 = select(sssgrd, 'value > %grd$zmin% and value < %grd$zmax%')
ssscov = gridpoly(sssout1)

chlout = selectpolygon(chlgrd, fishareas)
&describe chlout
chlout1 = select(chlgrd, 'value > %grd$zmin% and value < %grd$zmax%')
chlcov = gridpoly(chlout1)
q

intersect sstcov ssscov sstsss # 0.00000001
intersect sstsss chlcov efhcov # 0.00000001

ap
display 9999
mape catch
arcs catch
polygonshades efhcov 4
--------------------------------
```

# INDEX

T - #0418 - 071024 - C8 - 234/156/11 - PB - 9780367396138 - Gloss Lamination